The Stanford Mathematics Problem Book

With Hints and Solutions

George Polya
Jeremy Kilpatrick

9/09

Dover Publications, Inc.
Mineola, New York

Selections from the following works are reprinted by permission of the publishers:
How to Solve It: A New Aspect of Mathematical Method, by G. Polya.
Copyright 1945 by Princeton University Press, © 1957 by G. Polya;
    Princeton Paperback, 1971. Ref. [17].
Mathematics and Plausible Reasoning, by G. Polya.
    Vol. I: *Induction and Analogy in Mathematics.* Published 1954 by
    Princeton University Press. Ref. [18].
    Vol. II: *Patterns of Plausible Inference.* Rev. ed. with Appendix.
    Copyright © 1968 by Princeton University Press. Ref. [19].
Mathematical Discovery, by G. Polya. Copyright © 1962, 1965, 1968 by
    John Wiley and Sons, Inc. Vols. I, II. Refs. [20, 21].
California Mathematics Council Bulletin. Refs. [9–16].
American Mathematical Monthly. Ref. [24].

Copyright

Bibliographical Note

This Dover edition, first published in 2009, is an unabridged republication of the work originally published in 1974 by Teachers College Press,
New York.

Library of Congress Cataloging-in-Publication Data

Pólya, George, 1887–1985.
    The stanford mathematics problem book : with hints and solutions /
George Polya and Jeremy Kilpatrick. — Dover ed.
        p. cm.
    Includes bibliographical references.
    ISBN-13: 978-0-486-46924-9
    ISBN-10: 0-486-46924-7
    1. Mathematics—Problems, exercises, etc. I. Kilpatrick, Jeremy. II.
Title.

QA43.P63 2009
510.76—dc22

                                                    2008050086

        Manufactured in the United States of America
    Dover Publications, Inc., 31 East 2nd Street, Mineola, N.Y. 11501

Contents

Part One

INTRODUCTION

For twenty years, from 1946 to 1965, the Department of Mathematics at Stanford University conducted a competitive examination for high school seniors. The immediate and principal purpose of the examination was to identify, among each year's high school graduates, singularly capable students and attract them to Stanford. The broader purpose was to stimulate interest in mathematics among high school students and teachers generally, as well as the public.

The examination was modeled on the Eötvös Competition [see 23],* which was organized in Hungary in 1894 and which, in turn, appears to have been suggested by similar competitions in England and France. Gabor Szegö, chairman of the Stanford Department of Mathematics in 1946 and winner of the Eötvös Competition in 1912, initiated the Stanford examination.

The examination was established in the belief that an early manifestation of mathematical ability is a definite indication of exceptional intelligence and suitability for intellectual leadership in any field of endeavor. Furthermore, mathematical ability can be tested at a comparatively early age because it is manifested "not so much by the amount of accumulated knowledge as by the originality of mind displayed in the game of grappling with difficult though elementary problems [2, p. 406]."

As Buck [1] noted some years ago in reviewing mathematical competitions, an examination can be designed, broadly speaking, to test either achievement or aptitude. The Stanford University Competitive Examination in Mathematics was of the latter type. It emphasized

> originality and insight rather than routine competence . . . . A typical question might call for specific knowledge within the reach of those being tested, but would call for the employment of this in unusual ways requiring a high degree of ingenuity. The question may in fact introduce certain concepts which are quite unfamiliar to the student. In short, the winning student is asked to demonstrate research ability [1, pp. 204–205].

* Numbered references are on pages 67–68.

1

The first Stanford examination, in 1946, was administered in 60 California high schools to 322 participants. The winner was awarded a one-year scholarship by Stanford University; honorable mention and a mathematics book were given to three other participants. In 1953, the examination was extended beyond California to include Arizona, Oregon, and Washington; the number of scholarships was increased to two; and the number of honorable mention awards and books was increased to ten or so. From 1958 to 1962, the examination was co-sponsored by Sylvania Electric Products, Inc. The last examination, in 1965, was administered to about 1200 participants in 151 centers in California, Arizona, Idaho, Montana, Nevada, Oregon, and Washington. Cash prizes of $500, $250, and $250 were awarded to the three winners; honorable mention and a mathematics book went to eighteen participants. The examination was discontinued after 1965 mainly because the Stanford Department of Mathematics turned its interest to more graduate teaching.

Announcements of the examination were sent each year to all public and private high schools in each state where the examination was to be administered. Larger schools were designated as centers; students from other schools were free to arrange to take the examination in a convenient location.

The examination was administered by teachers and school personnel on a Saturday afternoon in March or April. The participants were given three hours to attempt three to five problems. The following instructions were, given:

> No books or notebooks may be used. You may not be able to do all the problems in three hours, but whatever you do should be carefully thought out. Scratch paper may be used. Either pen or pencil may be used. No questions concerning the test should be asked of the person in charge.
> Good presentation *counts!*
> It should be clear, concise, complete.

The papers were read in a two-stage process: First, they were read by teams of graduate students in the Department of Mathematics, including, as was sometimes possible, graduate students who were experienced high school teachers. Each team of two students was assigned a problem to read in as many papers as they could handle. Papers containing either a stated minimum of good solutions (for example, one and a half or two out of four) or some special feature were forwarded to the second stage. In the second stage, each paper that survived the first screening was read by at least one faculty member of the Department. The papers considered most likely to be winners were read by all participating faculty members.

To make the selection of winners easier, the problems were devised so that only a very few participants would be able to solve all of them.

On the other hand, to avoid too much frustration, the first problem was usually more accessible than the others, especially in the later years, so that many participants were able to solve it.

Although the mathematical content of the problems did not go beyond that of the high school curriculum, the problems were of types seldom found in textbooks. The purpose of such problems was not only to test the students' originality, but also to enrich the high school mathematics program by suggesting some new directions for students' and teachers' work. The types of problems included: (1) "guess and prove," in which one first guesses and then proves a mathematical fact; (2) "test consequences," in which one tests the consequences of a general statement; (3) "you may guess wrong," in which a highly plausible guess is incorrect; (4) "small scale theory," in which a sequence of subproblems illustrates theory construction; and (5) "red herring," in which an obvious relationship among the data turns out to be irrelevant to the solution [see 9; 19, pp. 160–161, ex. 1; 21, p. 139, ex. 14.23].

The problems were of the sort used as illustrations in *How to Solve It* [17], in the two volumes of *Mathematics and Plausible Reasoning* [18, 19], and in the two volumes of *Mathematical Discovery* [20, 21]. In fact, many of the problems appear, usually with solutions, in one or another of those books.

Most of the problems have also appeared in journals. The problems and the list of winners for each examination from 1946 to 1953 (except 1952) were published in the *American Mathematical Monthly* [2, 3, 4, 5, 6, 7, 8]; and the complete set of problems, together with an Introduction, somewhat adapted here, appears in the June–July 1973 issue of that monthly [24]. Articles containing problems, solutions, comments, and lists of winners for 1953 to 1961 (except 1959) were published in the *California Mathematics Council Bulletin* [9, 10, 11, 12, 13, 14, 15, 16].

The complete set of problems has never before, however, been published together with hints and solutions for all problems. Material copyrighted previously is reprinted here by permission.

Part Two contains the complete set of problems from the Stanford University Competitive Examination in Mathematics. They are numbered sequentially, by year and problem number, as follows: **46.1** designates problem 1 in the 1946 examination.

Part Three contains a hint for each problem; the hints are numbered to correspond with the problems. The hints are similar to those in Part IV of *How to Solve It* [17], and most of them use one or more of the heuristic questions and suggestions treated in that book.

Part Four contains a solution for each problem (sometimes two solutions); the numbering is as before. Solutions outline the procedures used, but some details are, of course, left to the reader. Some solutions end with an indication of connections to other problems or generalizations.

References to articles and books in which problems have appeared previously (with or without solutions) are given in brackets with the corresponding solutions in Part Four.

Many of the hints and solutions have come from discussions in seminars on problem solving held at Stanford University and at Teachers College. The problems have been used to illustrate problem-solving techniques with freshmen, prospective teachers, and experienced teachers alike. Teachers, and teachers of teachers, may find some useful suggestions on how to use the problems in *Mathematical Discovery* [20, pp. 209–212].

Part Two

PROBLEMS

46.1. In a tennis tournament there are $2n$ participants. In the first round of the tournament each participant plays just once, so there are n games, each occupying a pair of players. Show that the pairing for the first round can be arranged in exactly

$$1 \times 3 \times 5 \times 7 \times 9 \ldots \times (2n - 1)$$

different ways.

46.2. In a tetrahedron (which is not necessarily regular) two opposite edges have the same length a and they are perpendicular to each other. Moreover they are each perpendicular to a line of length b which joins their midpoints. Express the volume of the tetrahedron in terms of a and b, and prove your answer.

46.3. Consider the following four propositions, which are not necessarily true.

      I. If a polygon inscribed in a circle is equilateral it is also equiangular.

      II. If a polygon inscribed in a circle is equiangular it is also equilateral.

      III. If a polygon circumscribed about a circle is equilateral it is also equiangular.

      IV. If a polygon circumscribed about a circle is equiangular it is also equilateral.

(A) State which of the four propositions are true and which are false, giving a proof of your statement in each case.

(B) If, instead of general polygons, we should consider only quadrilaterals which of the four propositions are true and which are false? And if we consider only pentagons?

In answering (B) you may state conjectures, but prove as much as you can and separate clearly what is proved and what is not.

PROBLEMS 47

47.1. To number the pages of a bulky volume the printer used 1890 digits. How many pages has the volume?

47.2. Among grandfather's papers a bill was found:

72 turkeys $—67.9—

The first and last digit of the number that obviously represented the total price of those fowls are replaced here by blanks, for they have faded and are now illegible.

What are the two faded digits and what was the price of one turkey?

47.3. Determine m so that the equation in x

$$x^4 - (3m + 2)x^2 + m^2 = 0$$

has four real roots in arithmetic progression.

47.4. Let α, β, and γ denote the angles of a triangle. Show that

$$\sin \alpha + \sin \beta + \sin \gamma = 4 \cos \frac{\alpha}{2} \cos \frac{\beta}{2} \cos \frac{\gamma}{2},$$

$$\sin 2\alpha + \sin 2\beta + \sin 2\gamma = 4 \sin \alpha \sin \beta \sin \gamma$$

and

$$\sin 4\alpha + \sin 4\beta + \sin 4\gamma = -4 \sin 2\alpha \sin 2\beta \sin 2\gamma.$$

48.1 Consider the table:

$$1 = 1$$
$$2 + 3 + 4 = 1 + 8$$
$$5 + 6 + 7 + 8 + 9 = 8 + 27$$
$$10 + 11 + 12 + 13 + 14 + 15 + 16 = 27 + 64$$

Guess the general law suggested by these examples, express it in suitable mathematical notation, and prove it.

48.2. Three numbers are in arithmetic progression, three other numbers in geometric progression. Adding the corresponding terms of these two progressions successively, we obtain

85, 76, and 84

respectively, and adding all three terms of the arithmetic progression, we obtain 126. Find the terms of both progressions.

48.3. From the peak of a mountain, you see two points, A and B, in the plain. The lines of vision, directed to these points, include the angle γ. The inclination of the first line of vision to a horizontal plane is α, that of the second line β. It is known that the points A and B are on the same level and that the distance between them is c.

Express the elevation x of the peak above the common level of A and B in terms of the angles α, β, γ and the distance c.

48.4. A first sphere has the radius r_1. About this sphere circumscribe a regular tetrahedron. About this tetrahedron circumscribe a second sphere with radius r_2. About this second sphere circumscribe a cube. About this cube circumscribe a third sphere with radius r_3.

Find the ratios $r_1 : r_2 : r_3$ (which should be, according to Kepler, the ratios of the mean distances of the planets Mars, Jupiter, and Saturn from the Sun, but which are, in fact, rather different from the true ratios).

8

PROBLEMS 49

49.1. Prove that no number in the sequence

$$11, 111, 1111, 11111, \ldots$$

is the square of an integer.

49.2. The three sides of a triangle are of lengths l, m, and n, respectively. The numbers l, m, and n are positive integers,

$$l \leqq m \leqq n.$$

(A) Take $n = 9$ and find the number of different triangles of the described kind.

(B) Take various values of n and find a general law.

49.3. (A) Prove the following theorem: A point lies inside an equilateral triangle and has the distances x, y, and z from the three sides respectively; h is the altitude of the triangle. Then

$$x + y + z = h.$$

(B) State precisely and prove the analogous theorem in solid geometry concerning the distances of an inner point from the four faces of a regular tetrahedron.

(C) Generalize both theorems so that they should apply to any point in the plane or space, respectively (and not only to points inside the triangle or tetrahedron). Give precise statements and, if you have time, also proofs.

50.1. Observe that

$$1 = 1$$
$$1 - 4 = -(1 + 2)$$
$$1 - 4 + 9 = 1 + 2 + 3$$
$$1 - 4 + 9 - 16 = -(1 + 2 + 3 + 4)$$

Guess the general law suggested by these examples, express it in suitable mathematical notation, and prove it.

50.2. Given a square. Find the locus of the points from which the square is seen under an angle (A) of 90° (B) of 45°. (Let P be a point outside the square, but in the same plane. The smallest angle with vertex P containing the square is the "angle under which the square is seen" from P.) Sketch clearly both loci, give a full description, and a proof.

50.3. Call "axis" of a solid a straight line joining two points of the surface of the solid and such that the solid, rotated about this line through an angle which is greater than 0° and less than 360° coincides with itself.

A cube has 13 different axes, which are of three different kinds. Describe clearly the location of these axes, find the angle of rotation associated with each. Assuming that the edge of the cube is of unit length, compute the arithmetic mean of the lengths of the 13 axes. Do not use tables and compute to two decimals.

PROBLEMS 51

51.1. The length of the perimeter of a right triangle is 60 inches and the length of the altitude perpendicular to the hypotenuse is 12 inches. Find the sides of the triangle.

51.2. A quadrilateral is cut into four triangles by its two diagonals. We call two of these triangles "opposite" if they have a common vertex but no common side. Prove the following statements:

(A) The product of the areas of two opposite triangles is equal to the product of the areas of the other two opposite triangles.

(B) The quadrilateral is a trapezoid if, and only if, there are two opposite triangles equal in area.

(C) The quadrilateral is a parallelogram if, and only if, all four triangles are equal in area.

51.3. We consider the frustum of a right circular cone. The plane that is parallel to the lower and upper bases of the frustum and at equal distance from both intersects the frustum in the "median circle." The frustum and a cylinder have the same altitude, and the median circle of the frustum is the base of the cylinder.

Which one of these two solids has the greater volume, the frustum or the cylinder? Prove your answer!

(A possible proof is by algebra: Express both volumes in terms of suitable data and transform their difference so that its sign becomes obvious.)

52.1. Prove the proposition: If a side of a triangle is less than the average (arithmetic mean) of the two other sides, the opposite angle is less than the average of the two other angles.

52.2. Consider the frustum of a right pyramid with square base. Call "midsection" the intersection of the frustum with a plane parallel to the base and the top and at the same distance from both. Call "intermediate rectangle" the rectangle of which one side is equal to a side of the base and the other side is equal to a side of the top.

Four different friends of yours agree that the volume of the frustum equals the altitude multiplied by a certain area, but they disagree and make four different proposals regarding this area:

    I. the midsection
    II. the average of the base and the top
    III. the average of the base, the top, and the midsection
    IV. the average of the base, the top, and the intermediate rectangle.

Let h be the altitude of the frustum, a the side of its base, and b the side of its top. Express each of the four proposed rules in mathematical notation, decide whether it is right or wrong, and prove your answer.

52.3. Prove that the only solution of the equation

$$x^2 + y^2 + z^2 = 2xyz$$

in integers x, y, and z is $x = y = z = 0$.

12

PROBLEMS 53

53.1. Bob has 10 pockets and 44 silver dollars. He wants to put his dollars into his pockets so distributed that each pocket contains a different number of dollars.
(A) Can he do so?
(B) Generalize the problem, considering p pockets and n dollars. The problem is the most interesting when

$$n = \frac{(p+1)\,(p-2)}{2}.$$

Why?

53.2. Observe that the value of

$$\frac{1}{2!} + \frac{2}{3!} + \frac{3}{4!} + \cdots + \frac{n}{(n+1)!}$$

is $1/2$, $5/6$, $23/24$ for $n = 1, 2, 3$, respectively, guess the general law (by observing more values if necessary) and prove your guess.

53.3. Find x, y, u, and v satisfying the system of four equations

$$
\begin{aligned}
x + 7y + 3v + 5u &= 16 \\
8x + 4y + 6v + 2u &= -16 \\
2x + 6y + 4v + 8u &= 16 \\
5x + 3y + 7v + u &= -16
\end{aligned}
$$

(This may look long and boring: look for a shortcut.)

53.4. The four points G, H, V, and U are (in this order) the four corners of a quadrilateral. A surveyor wants to find the length $UV = x$. He knows the length $GH = l$ and measures the four angles

$$\angle GUH = \alpha, \ \angle HUV = \beta, \ \angle UVG = \gamma, \ \angle GVH = \delta.$$

(A) Express x in terms of α, β, γ, δ, and l.
(B) Find some way to test the correctness of the result.
(C) If you had a clear plan to do (A) characterize it in one short sentence.

54.1. Consider the table

$$1 = 1$$
$$3 + 5 = 8$$
$$7 + 9 + 11 = 27$$
$$13 + 15 + 17 + 19 = 64$$
$$21 + 23 + 25 + 27 + 29 = 125$$

Guess the general law suggested by these examples, express it in suitable mathematical notation, and prove it.

54.2. The side of a regular hexagon is of length n (n is an integer). By equidistant parallels to its sides, the hexagon is divided into T equilateral triangles each of which has sides of length 1. Let V denote the number of vertices appearing in this division, and L the number of boundary lines of length 1. (A boundary line belongs to one or two triangles, a vertex to two or more triangles.)

When $n = 1$ (which is the simplest case), $T = 6$, $V = 7$, $L = 12$. Consider the general case and express T, V, and L in terms of n. (Guessing is good, proving is better.)

54.3. Show that it is impossible to find (real or complex) numbers a, b, c, A, B, and C such that the equation

$$x^2 + y^2 + z^2 = (ax + by + cz)(Ax + By + Cz)$$

holds identically for independently variable x, y, and z.

14

55.1. Bob wants a piece of land, exactly level, which has four boundary lines. Two boundary lines run exactly north-south, the two others exactly east-west, and each boundary line measures exactly 100 feet. Can Bob buy such a piece of land in the U.S.? State your reasons!

55.2. (A) Find three numbers p, q, and r so that the equation
$$x^4 + 4x^3 - 2x^2 - 12x + 9 = (px^2 + qx + r)^2$$
holds identically for variable x.

(B) This problem requires the "exact" extraction of a square root of a given polynomial of degree 4, which may be possible in the present case, yet usually it is not. Why not?

55.3. Bob, Peter, and Paul travel together. Peter and Paul are good hikers; each walks p miles per hour. Bob has a bad foot and drives a small car in which two people can ride, but not three; the car covers c miles per hour. The three friends adopted the following scheme: They start together, Paul rides in the car with Bob, Peter walks. After a while, Bob drops Paul who walks on; Bob returns to pick up Peter, and then Bob and Peter ride in the car till they overtake Paul. At this point, they change: Paul rides and Peter walks just as they started and the whole procedure is repeated as often as necessary.

(A) How much progress (how many miles) does the company make per hour?

(B) Through which fraction of the travel time does the car carry just one man?

(C) Check the extreme cases $p = 0$ and $p = c$.

55.4. The vertex of a pyramid opposite the base is called the *apex*.

(A) Let us call a pyramid "isosceles" if its apex is at the same distance from all *vertices* of the base. Adopting this definition, prove that the base of an isosceles pyramid is *inscribed* in a circle the center of which is the foot of the pyramid's altitude.

(B) Now let us call a pyramid "isosceles" if its apex is at the same (perpendicular) distance from all *sides* of the base. Adopting this definition (different from the foregoing) prove that the base of an isosceles pyramid is *circumscribed* about a circle the center of which is the foot of the pyramid's altitude.

56.1. Given a regular hexagon and a point in its plane. Draw a straight line through the given point that divides the given hexagon into two parts of equal area.

56.2. I say that you can pay 50 cents in exactly 50 different manners. (The "manner" depends on how many coins of each kind—cents, nickels, dimes, quarters, half dollars—you use.) In how many manners can you pay 25 cents? Am I right about 50 cents? Justify your answer as clearly as you can.

56.3. Construct a hexagon by adding to an arbitrarily given triangle $\triangle$ three exterior isosceles triangles each of which has an angle of 120° opposite to that side of $\triangle$ that forms its base. Show that those three vertices of the hexagon that are not vertices of the given $\triangle$ are the vertices of an *equilateral* triangle. (It is enough to express just one side s of the allegedly equilateral triangle in terms of the sides a, b, and c of $\triangle$, provided that this expression for s is symmetric in a, b, and c.)

56.4. Ten people are sitting around a round table. The sum of ten dollars is to be distributed among them according to the rule that each person receives one half of the sum that his two neighbors receive jointly. Is there just one way to distribute the money? Prove your answer.

PROBLEMS 57

57.1. Bob's stamp collection consists of three books. Two tenths of his stamps are in the first book, several sevenths in the second book, and there are 303 stamps in the third book. How many stamps has Bob? (Is the condition sufficient to determine the unknown?)

57.2. We call a vertex of a tetrahedron *trirectangular* if the three edges starting from it are perpendicular to each other. Given the areas A, B, and C of the three faces adjacent to the trirectangular vertex of a tetrahedron, find the area D of the fourth face, opposite to that vertex. (Which problem of plane geometry would you regard as analogous?)

57.3. Divide a given triangle by three straight cuts into seven pieces four of which are triangles (and the remaining three pentagons). One of the triangular pieces is included by the three cuts, each of the three other triangular pieces is included by a certain side of the given triangle and two cuts.

(A) Choose the three cuts so that the four triangular pieces turn out to be congruent. Describe your choice precisely and draw a clear figure.

(B) Which fraction of the area of the given triangle is the area of a triangular piece in the dissection that you chose?

(It may be advantageous to examine first a particular shape of the given triangle for which the solution is particularly easy.)

58.1. How old is the captain, how many children has he, and how long is his boat? Given the product 32118 of the three desired numbers (integers). The length of the boat is given in feet (is several feet), the captain has both sons and daughters, he has more years than children, but he is not yet one hundred years old. (Give reasons for your answer.)

58.2. Find x, y, u, and v satisfying the system of four equations:

$$x + y + u = 4$$
$$y + u + v = -5$$
$$u + v + x = 0$$
$$v + x + y = -8$$

(This may look long and boring: look for a shortcut.)

58.3. "In any triangle the sum of the three . . . is greater than the semi-perimeter."

Replace the dots . . . successively by

    i. altitudes

    ii. medians

    iii. bisectors (of the angles).

You obtain so three different assertions. Examine each assertion: is it true or false? Prove your answer!

58.4. Observe that the value of

$$1!1 + 2!2 + 3!3 + \ldots + n!n$$

is 1, 5, 23, 119 for $n = 1, 2, 3, 4$, respectively. Guess the general law (by observing more values if necessary) and prove your guess.

18

PROBLEMS 59

59.1. Al and Bill live at opposite ends of the same street. Al had to deliver a parcel at Bill's home, Bill one at Al's home. They started at the same moment, each walked at constant speed and returned home immediately after leaving the parcel at its destination. They met the first time at the distance of a yards from Al's home and the second time at the distance of b yards from Bill's home.

(A) How long is the street?

(B) If $a = 300$ and $b = 400$, who walks faster?

59.2. Pennies (equal circles) are arranged in a regular pattern all over a very-very large table (the infinite plane). We examine two patterns.

In the first pattern, each penny touches four other pennies and the straight lines joining the centers of the pennies in contact dissect the plane into equal squares.

In the second pattern, each penny touches six other pennies and the straight lines joining the centers of the pennies in contact dissect the plane into equal equilateral triangles.

Compute the percentage of the plane covered by pennies (circles) for each pattern.

59.3. Prove: If n is an integer greater than 1, $n^{n-1} - 1$ is divisible by $(n - 1)^2$.

59.4. Erect an (exterior) square on each side of an (arbitrarily given) triangle. Those 6 vertices of these 3 squares that do not coincide with a vertex of the triangle form a hexagon. Three sides of this hexagon are, of course, equal to the corresponding sides of the triangle. Show that each one of the remaining three sides equals the double of a median of the triangle.

60.1. A certain make of ball point pen was priced 50 cents in the store opposite the high school but found few buyers. When, however, the store had reduced the price, the whole remaining stock was sold for $31.93. What was the reduced price? (Is the condition sufficient to determine the unknown?)

60.2. The point P is so located in the interior of a rectangle that the distance of P from a corner of the rectangle is 5 yards, from the opposite corner 14 yards, and from a third corner 10 yards. What is the distance of P from the fourth corner?

60.3. Prove the identity

$$\cos \frac{\alpha}{2} \cos \frac{\alpha}{4} \cos \frac{\alpha}{8} = \frac{\sin \alpha}{8 \sin \frac{\alpha}{8}}$$

and generalize.

60.4. Of twelve congruent equilateral triangles eight are the faces of a regular octahedron and four the faces of a regular tetrahedron. Find the ratio of the volume of the octahedron to the volume of the tetrahedron.

20

61.1. Solve the following system of three equations for the unknowns x, y and z:

$$5732x + 2134y + 2134z = 7866,$$
$$2134x + 5732y + 2134z = 670,$$
$$2134x + 2134y + 5732z = 11464.$$

61.2. It was a very hot day and the 4 couples drank together 44 bottles of coca-cola. Ann had 2, Betty 3, Carol 4 and Dorothy 5 bottles. Mr. Brown drank just as many bottles as his wife, but each of the other men drank more than his wife: Mr. Green twice, Mr. White three times and Mr. Smith four times as many bottles. Tell the last names of the four ladies. (Prove your answer.)

61.3. Solve the following system of three equations for the unknowns x, y, and z (a, b, and c are given):

$$x^2y^2 + x^2z^2 = axyz,$$
$$y^2z^2 + y^2x^2 = bxyz,$$
$$z^2x^2 + z^2y^2 = cxyz.$$

61.4. A pyramid is called "regular" if its base is a regular polygon and the foot of its altitude is the center of its base. A regular pyramid has a hexagonal base the area of which is one quarter of the total surface-area S of the pyramid. The altitude of the pyramid is h. Express S in terms of h.

62.1. Solve the system

$$2x^2 - 4xy + 3y^2 = 36$$
$$3x^2 - 4xy + 2y^2 = 36$$

(One solution is easy to guess, but you are required to find *all* solutions. Knowledge of analytic geometry is not needed to solve this problem, but may help to understand the result—how?)

62.2. Each of the four numbers a, b, c, and d is positive and less than one. Show that not all four products

$$4a(1 - b), 4b(1 - c), 4c(1 - d), 4d(1 - a)$$

are greater than one.

62.3. On each side of a right triangle, erect an exterior square (as it is usually done to illustrate Pythagoras' theorem). Join the vertex of the triangle's right angle to the center of the square on the hypotenuse, and join the centers of the squares on the other two sides. Show that the two line segments so obtained are
   (A) perpendicular to each other and
   (B) of equal length.

62.4. Five edges of a tetrahedron are of the same length a, and the sixth edge is of the length b.
   (A) Express the radius of the sphere circumscribed about the tetrahedron in terms of a and b.
   (B) How would you use the result (A) to determine practically the radius of a spherical surface (of a lens)?

PROBLEMS 63

63.1. In a right triangle, c is the length of the hypotenuse, a and b are the lengths of the two other sides, and d is the length of the diameter of the inscribed circle. Prove that

$$a + b = c + d$$

63.2. Show that the expression

$$n^2(n^2 - 1)\ (n^2 - 4)$$

is divisible by 360 for $n = 1, 2, 3, \ldots$.

63.3. Solve the system of three equations for the unknowns x, y, and z, giving all solutions:

$$x^2 + 5y^2 + 6z^2 + 8(yz + zx + xy) = 36$$
$$6x^2 + \ y^2 + 5z^2 + 8(yz + zx + xy) = 36$$
$$5x^2 + 6y^2 + \ z^2 + 8(yz + zx + xy) = 36$$

(One solution is easy to find.)

63.4. The base of a right prism is a regular hexagon, and the height of the prism is equal to the diameter of the circle inscribed in the base. The volume of the prism is equal to the volume of a regular octahedron.

Find the ratio of the surface-areas of these two solids.

Observe that the two solids have the same number of faces, and one of them is a regular solid, but the other is not. Any remark?

64.1. A cake has the shape of a right prism with a square base; it has icing on the top as well as on the sides (that is, on the four lateral faces). The altitude of the prism is 5/16 of the side of its base. Cut the cake into 9 pieces so that each piece has the same amount of cake and the same amount of icing. One of the 9 pieces should be a right prism with a square base with icing only on the top: Compute the ratio of its altitude to a side of its base and give a clear description, with an acceptable sketch, of all 9 pieces.

64.2. Show that each number of the sequence

$$49, 4489, 444889, 44448889, \ldots$$

is a perfect square.

64.3. If the area of a triangle is rational (that is, measured by a rational number) there are four thinkable cases: The triangle may have three or two rational sides, or just one or no rational side. Show by (preferably simple) examples that all four cases are actually possible.

64.4. An examination in three subjects, Algebra, Biology, and Chemistry was taken by 41 students. The following table shows how many students failed in each single subject and in their various combinations:

in	A	B	C	AB	AC	BC	ABC
failed	12	5	8	2	6	3	1

(For instance, 5 students failed in Biology, among whom there were 3 failing both in Biology and in Chemistry, and just one of these 3 failed in all three subjects.)

How many students passed in all three subjects?

(Can you think of a suitable diagram that would clarify the underlying idea?)

64.5. Let a, b, and c denote the lengths of the sides of a triangle, and d the length of the bisector of the angle opposite to the side of length c, terminated on the side.

(A) Express d in terms of a, b, and c.

(B) Check the expression obtained in as many ways as you can (by particular cases, limiting cases, and so on).

24

PROBLEMS 65

65.1. "How many children have you, and how old are they?" asked the guest, a mathematics teacher.

"I have three boys," said Mr. Smith. "The product of their ages is 72 and the sum of their ages is the street number."

The guest went to look at the entrance, came back and said: "The problem is indeterminate."

"Yes, that is so," said Mr. Smith, "but I still hope that the oldest boy will some day win the Stanford competition."

Tell the ages of the boys, stating your reasons.

65.2. Of a right triangle, given the length of the hypotenuse c and the area A. On each side of the triangle, describe a square exterior to the triangle and consider the least convex figure containing the three squares (formed by a tight rubber band around them): it is a hexagon (which is irregular, has one side in common with each square, and one of its remaining three sides is obviously of length c).

Find the area of the hexagon.

65.3. Let the numbers, x, y, and 1 measure the lengths of the three sides of some triangle and suppose that

$$x \leqq y \leqq 1.$$

Let the point (x, y), with rectangular coordinates x and y, *represent* the triangle on a plane. Describe precisely and sketch clearly the set of those points of the plane that, in the manner explained, represent

(A) triangles,
(B) isosceles triangles,
(C) right triangles,
(D) acute triangles,
(E) obtuse triangles.

Locate the representative points of still other noteworthy triangular shapes.

65.4. Find the remainder of the division of the polynomial

$$x + x^9 + x^{25} + x^{49} + x^{81}$$

by the polynomial $x^3 - x$.

Part Three

HINTS

46.1. *Could you restate the problem?* Imagine that you are one of the participants. In how many ways can you be paired with someone else? Do you see how the problem can be divided into two parts: (A) choosing your antagonist, and (B) pairing the remaining players?

46.2. *Look at the unknown.* The unknown is the volume of a tetrahedron, which can be computed when the base and the height are given. But neither the base nor the height is given in this problem. *Could you imagine a more accessible related problem?* (Do you see a more accessible tetrahedron that is an aliquot part of the original one?)

46.3. Proving any one of the propositions reduces to proving either line segments or angles equal. Do you know a theorem or theorems useful in proving such things equal? *Draw a figure.* Could you introduce some auxiliary elements to make possible the use of the theorems you have recalled?

47.1. *Here is a problem related to yours:* If the book has exactly 9 numbered pages, how many digits does the printer use? (Answer: 9.) Here is another *problem related to yours:* If the book has exactly 99 numbered pages, how many digits does the printer use?

47.2. *Could you restate the problem?* What can the two faded digits be if the total price, expressed in cents, is divisible by 72?

47.3. *What is the condition?* The four roots must form an arithmetic progression. Yet the equation has a particular feature: it contains only even powers of the unknown x. Therefore, if a is a root, $-a$ is also a root.

47.4. *Do you know a related theorem?* Notice the similarity between the three identities, especially in their left-hand sides. If you had established one identity, how could you derive the other two?

HINTS 48

48.1. Discovery by induction needs observation. Observe the terms on the right-hand sides, the initial terms of the left-hand sides, and the final terms. What is the general law?

48.2. *Separate the various parts of the condition. Can you write them down?* Let

$$a - d, \qquad a, \qquad a + d$$

be the terms of the arithmetic progression, and

$$bg^{-1}, \qquad b, \qquad bg$$

be the terms of the geometric progression.

48.3. *Separate the various parts of the condition. Can you write them down?* Let a and b stand for the lengths of the (unknown) lines of vision, α and β for their inclinations to the horizontal plane, respectively. We may distinguish three parts in the condition, concerning

(1) the inclination of a
(2) the inclination of b
(3) the triangle with sides a, b, and c.

48.4. *Look at the unknown.* There are two unknowns: the ratio $r_1 : r_2$ and the ratio $r_2 : r_3$.

HINTS 49

49.1. *Could you restate the problem?* As a "problem to find," it becomes: "Find a perfect square s of the form $1 + 10 + 10^2 + \ldots + 10^k$, where k is a positive integer." *Separate the various parts of the condition. Can you write them down?* We may distinguish two parts in the condition:

(1) s must be a square
(2) s must have the desired form.

49.2. Discovery by induction needs observation. Can you find a systematic way of counting the triangles for a given value of n?

49.3. *Do you know a related theorem?* The altitude is h. Do you know a simpler theorem concerning the altitude of a triangle?

50.1. Discovery by induction needs observation. Examine the transition from one case to the next.

50.2. *Do you know a related problem?* The locus of the points from which a given segment of a straight line is seen under a given angle consists of two circular arcs, ending in the extreme points of the segment, and symmetric to each other with respect to the segment.

50.3. Certain axes are easily found just by inspection—but are they *all* the axes? *Can you prove* that your list of axes is exhaustive? Has your list a clear principle of classification?

51.1. *Separate the various parts of the condition. Can you write them down?* We may distinguish three parts in the condition, concerning
  (1) perimeter
  (2) right triangle
  (3) height to hypotenuse.

51.2. *Draw a figure. Introduce suitable notation.* How can you show that areas (or products of areas) are equal?

51.3. *Draw a figure. Introduce suitable notation.*

52.1. What is the hypothesis? What is the conclusion? Let a, b, and c denote the sides, and A, B, and C the opposite angles, respectively. Then the hypothesis is that

$$a < \frac{b + c}{2}$$

and the conclusion is that

$$A < \frac{B + C}{2}$$

Look at the conclusion. Could you restate it?

52.2. *Do you know a related problem? A more special problem?* What happens when you vary the data of the problem?

52.3. *What is the condition?* The sum of the squares of three integers must be twice their product. The sum must be an even number.

HINTS 53

53.1. If Bob had very many dollars, he would have obviously no difficulty in filling each of his pockets differently. *Could you restate the problem?* What is the minimum number of dollars that can be put in 10 pockets so that no two different pockets contain the same amount?

53.2. Do you *recognize* the denominators 2, 6, 24? *Do you know a related problem? An analogous problem?*

53.3. To solve such a system we have to combine the equations in some way—look out for relations between the equations which could indicate a particularly advantageous combination.

53.4. Newton once observed that in certain geometric problems one obtains the same system of equations regardless of which quantities are considered as the data and which are considered as the unknowns. Consequently, one should choose the data and unknowns so that it is easy to set up the equations.

HINTS 54

54.1. Discovery by induction needs observation. Observe the right-hand sides, the initial terms of the left-hand sides, and the final terms. What is the general law?

54.2. *Draw a figure.* It may help you discover the law inductively, or it may lead you to relations between T, V, L, and n.

54.3. *Is it possible to satisfy the condition? Is the condition sufficient to determine the unknown?* How could you split the condition into appropriate parts?

HINTS 55

55.1. *What is the question?* What does it mean? It is, in fact, a question of interpretation: You are supposed to interpret "level," "east," "west," "north," and "south" on an *idealized*, exactly spherical globe.

55.2. *Is it possible to satisfy the condition? Is the condition sufficient to determine the unknown?* How could you split the condition into appropriate parts?

55.3. *Separate the various parts of the condition. Can you write them down?* Between the start and the point where the three friends meet again there are three different phases:

      (1) Bob rides with Paul

      (2) Bob rides alone

      (3) Bob rides with Peter.

Call t_1, t_2, and t_3 the duration of these phases, respectively. How could you split the condition into appropriate parts?

55.4. *Do you know a related theorem?* The foot of the altitude is the midpoint of the base in an isosceles triangle. This is a *theorem related to yours and proved before. Could you use its method?* The theorem on the isosceles triangle is proved from congruent right triangles of which the altitude is a common side.

HINTS 56

56.1. *Could you imagine a more accessible related problem? A more general problem?*

56.2. *Could you imagine a more accessible related problem? A more general problem? An analogous problem?* Here is a *very* simple analogous problem: In how many ways can you pay one cent? Here is a more general problem: In how many ways can you pay an amount of n cents using cents, nickels, dimes, quarters, and half-dollars? This problem can be solved by inspection for simple particular cases, as shown in the short table below (where E_n denotes the number of different ways of paying an amount of n cents using the five kinds of coins).

n	4	5	9	10	14	15	19	20	24
E_n	1	2	2	4	4	6	6	9	9

We are especially concerned with E_{25} and E_{50}, but our question is general (to compute E_n for general n)—yet it is still "isolated." Here is a *very* simple analogous problem: Find A_n, the number of ways to pay an amount of n cents using only cents. ($A_n = 1$.)

56.3. *Draw a figure. Introduce suitable notation.* How can you obtain an expression for s? By Euclidean methods? By analytic geometry? By trigonometry? Which appears most promising?

56.4. *Introduce suitable notation. What is the condition?* How is each person's share related to that of his neighbors? To that of his left-hand neighbor? Who gets the maximum amount of money?

HINTS 57

57.1. *What is the unknown? What are the data? What is the condition?*

57.2. *What is the unknown?* The area, D, of a triangle. How can you get this kind of thing? The area of a triangle can be computed by Heron's formula if the three sides are known. Let a, b, and c denote the lengths of the sides, and let $s = (a + b + c)/2$; then

$$D^2 = s(s - a)(s - b)(s - c).$$

The sides a, b, and c are in the right triangles whose areas are A, B, and C, respectively. Let the legs of these triangles have lengths p, q, and r, so that

$$a^2 = q^2 + r^2, \qquad b^2 = r^2 + p^2, \qquad c^2 = p^2 + q^2.$$

But the areas A, B, and C are given by

$$A = \frac{1}{2} qr, \qquad B = \frac{1}{2} rp, \qquad C = \frac{1}{2} pq.$$

We have seven unknowns—D, a, b, c, p, q, r—and a system of seven equations to determine them. Yet there is a snag: Solving the system appears to be a lot of trouble, and Heron's formula may not look too inviting. Let's try a new start.

What is the unknown? The area, D, of a triangle. How can you get this kind of thing? The most familiar way to compute the area of a triangle is

$$D = \frac{ah}{2},$$

where a is the base, and h the altitude, of the triangle with area D. Let a have the same meaning as above and let us introduce h into the figure.

57.3. *Could you imagine a more accessible related problem? A more special problem?*

HINTS 58

58.1. *What is the unknown?* Let x, y, and z represent the number of children, the captain's age, and the length of the boat, respectively.

We may conceive the problem thus: the unknown is a triplet (x, y, z) of numbers.

Separate the various parts of the condition. Can you write them down?

      (A) x, y, and z are positive integers different from 1 and such that $xyz = 32118$.

      (B) $4 \leqq x < y < 100$.

Which clause, (A) or (B), is more manageable?

58.2. To solve such a system we have to combine the equations in some way. Observe that any permutation of x, y, u, and v leaves the system of the left-hand sides unchanged. This *symmetry* suggests that we should treat all four equations symmetrically.

58.3. *Could you imagine a more accessible related problem? A more general problem?* What is common to all three assertions? Could you make a generalization?

    Each assertion refers to an inequality. Do you know a theorem that could be useful?

58.4. Discovery by induction needs observation. Can you see a pattern in the cases observed?

HINTS 59

59.1. *What is the unknown? What are the data? What is the condition?* Can you write an equation that expresses a part of the condition? Recall that an equation expresses the same quantity in two different ways.

59.2. *Draw a figure.* What is the relationship between the percentage of the plane covered by pennies and the percentage of each equal square or equal equilateral triangle covered by pennies?

59.3. What is the hypothesis? What is the conclusion? The hypothesis is that n is an integer greater than 1. The conclusion is that $n^{n-1} - 1 = (n-1)^2 P(n)$, where $P(n)$ denotes an integer depending on n.

59.4. *Draw a figure. Introduce suitable notation.* How can you show that line segments are in the ratio 1 : 2 ?

HINTS 60

60.1. *What is the unknown? What are the data? What is the condition?*

60.2. *Could you imagine a more accessible related problem? A more general problem?* Here is a more general problem: The point P lies in the interior of a rectangle, its distances from the four corners are a, b, c, and d, in cyclical order (as they are met by the hands of a watch). Find d in terms of a, b, and c.

60.3. *Do you know a related theorem?* Try to think of a simpler analogous identity.

60.4. *What is the unknown?* The unknown is the ratio of the volume of a regular octahedron to the volume of a regular tetrahedron. What do the octahedron and the tetrahedron have in common? Their faces are congruent equilateral triangles; their edges are equal. *Could you restate the problem?* Compute the volume of a regular octahedron and the volume of a regular tetrahedron given the length of an edge.

    Draw a figure. Introduce suitable notation. The solution to a problem in solid geometry often depends on a "key plane figure" that opens the door to essential relations.

HINTS 61

61.1. To solve such a system we have to combine the equations in some way. The system of the three left-hand sides is *symmetric* with respect to x, y, and z; that is, it remains unchanged under any permutation of x, y, and z. This suggests that we should treat all three equations symmetrically.

61.2. Obviously, the number of possibilities is restricted from the start $(4! = 24)$. Yet, if you are smart, you need not examine all these cases. *Separate the various parts of the condition. Can you write them down?* Let b, g, w, and s stand for the number of bottles consumed by the wife of Brown, Green, White, and Smith, respectively.

61.3. To solve such a system we have to combine the equations in some way. The symmetry of the system of the three left-hand sides suggests that we treat all three equations symmetrically. A natural first step is to eliminate the product xyz on the right-hand sides by dividing, but this forces the examination of those cases in which xyz vanishes.

61.4. *Draw a figure. Introduce suitable notation.* The solution to a problem in solid geometry often depends on a "key plane figure."

62.1. Observe that the system is *symmetric* with respect to the two unknowns x and y. This symmetry should somehow be mirrored in the solution.

62.2. Observe that the set of four products is *symmetric* with respect to a, b, c, and d. This symmetry should somehow be mirrored in the solution.

62.3. *Draw a figure. Introduce suitable notation.* Denote the vertices of the triangle by A, B, and C, where C is the vertex of the right angle. Let A', B', and C' be the centers of the squares described on sides BC, CA, and AB, respectively. In the figure, C appears to lie on $A'B'$. Do you see any advantage in attempting to prove this conjecture?

62.4. *Draw a figure. Introduce suitable notation.* The solution to a problem in solid geometry often depends on a "key plane figure."

63.1. *Go back to definitions.* What is the inscribed circle of a triangle? *Do you know a theorem that could be useful?*

63.2. *What is the conclusion?* The expression must be divisible by 360. Yet the expression has an unusual feature: its three factors can themselves be factored, to yield a product of six factors.

63.3. Observe that the system is *symmetric* with respect to the three unknowns x, y, and z. This symmetry should somehow be mirrored in the solution.

63.4. Since the two solids have the same number of faces and the same volume, try to guess which one has the smaller surface area. The correctness of your guess is less important than how you test it.

64.1. *What is the unknown?* The ratio of the altitude of a right prism to a side of its square base. How can you get this kind of thing? If you calculate expressions for the two quantities, you can take their ratio. What are these quantities? How can you get expressions for them?

64.2. Consider the general case: what is the *n*th term of the sequence? Can you guess what its positive square root might be? How could you test this guess?

64.3. As long as you are free to choose your own example in each case, you might as well select a triangle whose area is easy to calculate.

64.4. What are the various possibilities for passing or failing the three subjects? Can you think of some systematic way of counting how many students fell into each category?

64.5. *Draw a figure. Introduce suitable notation.* The bisector *d* divides the side *c* into two segments whose lengths we shall denote by *p* and *q*, respectively; the first has a common endpoint with *a*, the second with *b*. Now we have

three data: *a*, *b*, and *c*

three unknowns: *p*, *q*, and *d*

three triangles: the original one and the two smaller ones into which it is divided by *d*.

HINTS 65

65.1. *Separate the various parts of the condition. Can you write them down?* We may distinguish four parts in the condition, concerning
(1) the product of the boys' ages
(2) the sum of their ages
(3) the indeterminacy of the problem
(4) the fact (whose revelance may pass unnoticed) that one of the boys is *the* oldest.

65.2. *What is the unknown?* The area of the hexagon. How do you get this kind of thing? *Is the condition sufficient to determine the unknown?*

65.3. *What is the unknown?* There are five unknowns: A set of points (A) and four subsets of it. How can you characterize such a set? By specifying the coordinates *x,y* of the points belonging to it—by relations between these coordinates!

65.4. *What is the unknown?* The remainder after division of one polynomial by another. *Could you restate the problem?* Denote the quotient by $q(x)$ and the remainder by $r(x)$. We are to find a polynomial $r(x)$ of degree not higher than 2 such that

$$x + x^9 + x^{25} + x^{49} + x^{81} = q(x)(x^3 - x) + r(x).$$

Part Four

SOLUTIONS

Note: References in brackets at the beginning of a solution are to books and articles in which the corresponding problem has appeared previously (in many cases with the solution given here or one very similar). The *American Mathematical Monthly* article containing the complete set of problems [24] is not included in these references. Page numbers refer to the statement of the problem itself in a given reference. "Cf." refers to different but related problems, or to cases in which a problem or solution varies somewhat in wording or content from the one given in this book. The interested reader will find it helpful to recall that only references [9–21] contain solutions, hints, or comments. All references are identified on pages 67–68.

46.1. [2; 18, p. 118]   Call the required number of pairings of $2n$ players P_n. If you are a participant, you can be matched with any one of the other $2n - 1$ players. Once your antagonist is chosen, there remain

$$2n - 2 = 2(n - 1)$$

players, who can be paired in P_{n-1} ways. Hence

$$P_n = (2n - 1)P_{n-1}.$$

46.2. [2; 17, p. 235; Cf. 20, pp. 109–110, ex. 4.17]   The plane through one edge of length a and the perpendicular of length b divides the tetrahedron into two *more accessible* congruent tetrahedra, each with base $ab/2$ and height $a/2$. Hence the required volume

$$= 2 \cdot \frac{1}{3} \cdot \frac{ab}{2} \cdot \frac{a}{2} = \frac{a^2 b}{6}.$$

46.3. [2; Cf. 19, pp. 161–162, ex. 4]

    (A) Propositions I and IV are generally true, but Propositions II and III are false.

    (B) Propositions II and III are false: the rectangle and the rhombus are counterexamples, respectively. Propositions II and III are true for pentagons, and follow from Propositions II′ and III′ respectively:

        II′. If a polygon inscribed in a circle is equiangular, any two sides separated by just one intervening side are equal.

        III′. If a polygon circumscribed about a circle is equilateral, any two angles separated by just one intervening angle are equal.

    To prove Propositions I, II′, III′, and IV, join the center of the circle to the vertices of the polygon, draw perpendiculars from the center to the sides, and pick out congruent triangles.

SOLUTIONS 47

47.1. [3; Cf. 17, p. 234, prob. 4] A volume of 999 pages needs

$$9 + 2 \times 90 + 3 \times 900 = 2889$$

digits. If the bulky volume in question has x pages,

$$189 + 3(x - 99) = 1890$$
$$x = 666.$$

47.2 [3; 17, p. 234] If $-679-$ is divisible by 72, it is divisible both by 8 and by 9. If it is divisible by 8, the number $79-$ must be divisible by 8 (since 1000 is divisible by 8) and so $79-$ must be 792: the last faded digit is 2. If -6792 is divisible by 9, the sum of its digits must be divisible by 9 (the rule about "casting out nines") and so the first faded digit must be 3. The price of one turkey was (in grandfather's time) $367.92 \div 72 = 5.11$.

47.3. [3; 17, p. 236] If a and $-a$ are the roots with least absolute value, the progression will be of the form

$$-3a, \ -a, \ a, \ 3a.$$

Hence

$$(x^2 - a^2)(x^2 - 9a^2) = x^4 - (3m + 2)x^2 + m^2.$$

Comparing coefficients of like powers, we obtain the system

$$10a^2 = 3m + 2$$
$$9a^4 = m^2.$$

Elimination of a yields

$$19m^2 - 108m - 36 = 0$$

Hence $m = 6$ or $-6/19$.

47.4. [3; 19, p. 162] Call the three identities (a), (b), and (c), respectively. If

$$\alpha + \beta + \gamma = \pi,$$

then

$$(\pi - 2\alpha) + (\pi - 2\beta) + (\pi - 2\gamma) = \pi.$$

We can pass from (a) to (b), and also from (b) to (c), by substituting $\pi - 2\alpha$, $\pi - 2\beta$, and $\pi - 2\gamma$ for α, β, and γ, respectively. It remains to verify (a), which can be done in many ways. For instance, substitute $2u$, $2v$, and $\pi - 2u - 2v$ for α, β, and γ, respectively. Then (a) becomes

$$\sin u \cos u + \sin v \cos v$$
$$= [2 \cos u \cos v - \cos (u + v)] \sin (u + v).$$

Use the addition theorems of cosine and sine.

SOLUTIONS 48

48.1. [4; 18, p. 8] The general law is:

$$(n^2 + 1) + (n^2 + 2) + \ldots + (n + 1)^2 = n^3 + (n + 1)^3$$

The terms on the left-hand side are in arithmetic progression.

48.2. [4; 17, p. 236] The condition is easily split into four parts expressed by the four equations

$$\begin{aligned} a - d + bg^{-1} &= 85 \\ a + b &= 76 \\ a + d + bg &= 84 \\ 3a &= 126. \end{aligned}$$

The last equation yields $a = 42$, then the second $b = 34$. Adding the remaining two equations, we obtain

$$2a + b(g^{-1} + g) = 169.$$

Since a and b are already known, we have here a quadratic equation for g. It yields

$$g = 2, d = -26 \quad \text{or} \quad g = \frac{1}{2}, d = 25.$$

The progressions are

68, 42, 16		17, 42, 67
	or	
17, 34, 68		68, 34, 17

48.3. [4; 17, p. 237] The three parts of the condition are expressed by

$$\sin \alpha = \frac{x}{a}$$

$$\sin \beta = \frac{x}{b}$$

$$c^2 = a^2 + b^2 - 2ab \cos \gamma.$$

The elimination of a and b yields

$$x^2 = \frac{c^2 \sin^2 \alpha \sin^2 \beta}{\sin^2 \alpha + \sin^2 \beta - 2 \sin \alpha \sin \beta \cos \gamma}.$$

48.4. [4; Cf. 18, p. 203, ex. 17] In the tetrahedron, an altitude is one leg of a right triangle whose hypotenuse is an edge. If the edge has length a, the other leg of the triangle has length $a/\sqrt{3}$ (it is two-thirds of the altitude of a face). Since the length of the altitude is $r_1 + r_2$, we obtain

$$r_1 + r_2 = \frac{a\sqrt{6}}{3}.$$

The center of the tetrahedron lies on the altitude, and the line segment of length r_2 joining the center to the opposite vertex of the right triangle is the hypotenuse of a second right triangle whose legs are of lengths r_1 and $a/\sqrt{3}$. Thus

$$r_2{}^2 - r_1{}^2 = \frac{a^2}{3}.$$

Dividing by the previous equation, we obtain the system

$$r_2 - r_1 = \frac{a\sqrt{6}}{6}$$

$$r_2 + r_1 = \frac{a\sqrt{6}}{3}$$

whose solution is $r_1 = a\sqrt{6}/12$ and $r_2 = a\sqrt{6}/4$. Hence $r_1 : r_2 = 1 : 3$.

In the cube, if b is the length of an edge, then $r_2 = b/2$ and $r_3 = b\sqrt{3}/2$. Hence $r_2 : r_3 = 1 : \sqrt{3}$ and $r_1 : r_2 : r_3 = 1 : 3 : 3\sqrt{3}$.

49.1. [5; 21, p. 192]   If s is a number in the sequence, s must have the form

$$11 + 100m = 4(25m + 2) + 3,$$

where m is a non-negative integer, and therefore s leaves a remainder of 3 when divided by 4. But squares are of the form $4n^2$ or $4n^2 + 4n + 1$ and hence leave remainders of either 0 or 1 when divided by 4.

49.2. [5; Cf. 18, p. 9, ex. 5]

(A) The values of l from 1 to 9 inclusive correspond to 1, 2, 3, 4, 5, 4, 3, 2, 1 triangles, respectively, or $5^2 = 25$ in all.

(B) A general law would be

$$\left(\frac{n+1}{2}\right)^2 \quad \text{or} \quad \left(\frac{n+1}{2}\right)^2 - \frac{1}{4},$$

according as n is odd or even. A uniform law for both cases: the integer nearest to $(n+1)^2/4$.

49.3. [5; 19, p. 161]

(A) Let a be a side of the equilateral triangle. Joining the point inside the triangle to its three vertices, you divide it into three triangles with areas that added together give the whole area: $ax/2 + ay/2 + az/2 = ah/2$. Divide by $a/2$.

(B) A point inside a regular tetrahedron with altitude h has the distances x, y, z, and w from the four faces, respectively. Then $x + y + z + w = h$. The proof is analogous: divide the regular tetrahedron into four tetrahedra.

(C) The relation remains valid in both cases (A) and (B) for outside points, provided that the distances x, y, z (and w) are taken with the proper sign: plus $(+)$ when a spectator placed in the point sees the side (face) from inside, minus $(-)$ when he sees it from outside. The proof is essentially the same.

SOLUTIONS 50

50.1. [6; 18, p. 116]   The general law is:

$$1 - 4 + 9 - 16 + \ldots + (-1)^{n-1} n^2 = (-1)^{n-1} \frac{n(n+1)}{2}.$$

The step from n to $n+1$ requires us to verify that

$$(-1)^n (n+1)^2 = (-1)^n \frac{(n+1)(n+2)}{2} - (-1)^{n-1}\frac{n(n+1)}{2}.$$

50.2. [6; Cf. 17, pp. 234–235, prob. 7] In any position the two sides of the angle must pass through two vertices of the square. As long as they pass through the same pair of vertices, the angle's vertex moves along the same circular arc (by the theorem underlying the hint). Hence each of the two loci required consists of several circular arcs: of 4 semicircles in the case (A) and of 8 quarter-circles in the case (B). (See [17, p. 244] for the figure.)

50.3. [6; Cf. 17, p. 235, prob. 8] The 13 axes may be classified as follows:
  - (1) 4 axes, each through two opposite vertices; angles 120°, 240°
  - (2) 6 axes, each through the midpoints of two opposite edges; angle 180°
  - (3) 3 axes, each through the center of two opposite faces; angles 90°, 180°, 270°.

The lengths are $\sqrt{3}$, $\sqrt{2}$, and 1, respectively, and the average is

$$\frac{4\sqrt{3} + 6\sqrt{2} + 3}{13} = 1.42.$$

SOLUTIONS 51

51.1. [7; 17, p. 237] Let a, b, and c denote the sides, the last being the hypotenuse. The three parts of the condition are expressed by

$$a + b + c = 60$$
$$a^2 + b^2 = c^2$$
$$ab = 12c.$$

Observing that

$$(a + b)^2 = a^2 + b^2 + 2ab$$

we obtain

$$(60 - c)^2 = c^2 + 24c.$$

Hence $c = 25$ and either $a = 15$, $b = 20$ or $a = 20$, $b = 15$ (no difference for the triangle).

51.2. [7; 19, p. 161] The quadrilateral has to be convex. Let us call I, II, III, IV the triangles into which it is divided by its diagonals; (I), (II), (III), (IV) the areas of the four triangles, respectively; and p, q, r, s the lengths of the four line segments from the vertices to the intersection of the diagonals. Name and number in "cyclic order" so that the side of length p is common to IV and I, q to I and II, r to II and III, s to III and IV; I is opposite to III, II to IV; $p + r$ is the length of one diagonal, $q + s$ that of the other. Let p and q include the angle α. Then

$$2(\text{I}) = pq \sin \alpha \qquad 2(\text{II}) = qr \sin \alpha$$
$$2(\text{III}) = rs \sin \alpha \qquad 2(\text{IV}) = sp \sin \alpha.$$

Hence

(A) $(\text{I})(\text{III}) = (\text{II})(\text{IV})$.

(B) The base of I is parallel to that of III if, and only if,

$$p/q = r/s \quad \text{or} \quad (\text{II}) = (\text{IV}).$$

(C) The quadrilateral is a parallelogram if, and only if,

$$p = r, q = s \quad \text{or} \quad (\text{I}) = (\text{II}) = (\text{III}) = (\text{IV}).$$

51.3. [7; Cf. 20, p. 51, ex. 2.54] Let h be the common altitude and a and b be the radii of the lower and upper base of the frustum, respectively. Then the radius of the cylinder is $(a + b)/2$. The difference of the volumes, frustum minus cylinder,

$$\pi h \left[\frac{a^2 + ab + b^2}{3} - \left(\frac{a + b}{2} \right)^2 \right] = \frac{\pi h (a - b)^2}{12}$$

is positive unless $a = b$ and the solids coincide.

SOLUTIONS 52

52.1. Since $A + B + C = 180°$, proving that $A < (B + C)/2$ amounts to proving that $A < (180° - A)/2$ or $A < 60°$. But $A < 60°$ is equivalent to $\cos A > 1/2$, which suggests the use of the law of cosines.

By hypothesis, $b + c > 2a$. Squaring both sides and applying the law of cosines, we obtain

$$b^2 + 2bc + c^2 > 4(b^2 + c^2 - 2bc \cos A)$$

or

$$8bc \cos A > 3b^2 + 3c^2 - 2bc.$$

Subtracting $4bc$ from both sides, we obtain

$$4bc(2 \cos A - 1) > 3(b - c)^2 \geqq 0.$$

Therefore $\cos A > 1/2$.

52.2. [19, p. 162; Cf. 17, pp. 210–211, and 21, pp. 2–7] According to the four proposals, the volume of the frustum would be, respectively,

$$\text{I.} \quad [(a + b)/2]^2 \, h$$
$$\text{II.} \quad [(a^2 + b^2)/2] \, h$$

          III. $[a^2 + b^2 + (a + b)^2/4]\, h/3$

          IV. $[a^2 + b^2 + ab]\, h/3.$

If $b = a$, the frustum becomes a prism with volume a^2h: all four proposals agree in yielding the correct result. If $b = 0$, the frustum becomes a pyramid with volume $a^2h/3$: only IV yields this, and so the others must be incorrect.

To prove that IV is generally correct, let x be the altitude of the small pyramid cut from the full pyramid to leave the frustum. If the volumes of the frustum, the full pyramid, and the small pyramid are F, A, and B, respectively, then

$$F = A - B = \frac{a^2(x + h)}{3} - \frac{b^2 x}{3} = [a^2 h + (a^2 - b^2)x]\, \frac{1}{3}.$$

A plane section of the figure through the altitude and parallel to one side of the base contains similar triangles whose sides yield the proportion

$$\frac{x}{x + h} = \frac{b}{a}$$

so that

$$x = \frac{bh}{a - b}.$$

Substituting this value for x in the expression for F, we obtain

$$F = [a^2 h + \frac{a^2 - b^2}{a - b}\, bh]\, \frac{1}{3} = [a^2 + b^2 + ab]\, \frac{h}{3}.$$

52.3.   Suppose x, y, and z are integers. Let 2^k, with $k \geqq 0$, be the highest power of 2 that divides x, y, and z, so that $x = 2^k x'$, $y = 2^k y'$, and $z = 2^k z'$. Then substituting in the given equation and dividing through by 2^{2k}, we obtain

$$(x')^2 + (y')^2 + (z')^2 = 2^{k+1} x' y' z'.$$

Since the right-hand side is even, so is the left-hand side, and either x', y', and z' are all even or just one of them is. But if x', y', and z' are not all zero (and if one is, the others are), they cannot all be even, because 2 is not a common factor. Suppose x' is even and y' and z' are odd. Subtracting $(x')^2$ from both sides of the above equation yields

$$(y')^2 + (z')^2 = x'(2^{k+1} y' z' - x').$$

Both $(y')^2$ and $(z')^2$ are of the form $4n^2 + 4n + 1$, and so the left-hand side divided by 4 leaves the remainder 2, whereas the right-hand side is divisible by 4 (both x' and the quantity in parenthesis are even): Contradiction.

53.1. [8; 9; Cf. 17, p. 234, prob. 3]

(A) The least possible number of dollars in a pocket is obviously 0. The next greater number is at least 1, the next greater at least 2, . . . , and the number in the last (tenth) pocket is at least 9. Therefore, the number of dollars required is at least

$$0 + 1 + 2 + 3 + \cdots + 9 = 45.$$

Bob cannot make it: he has only 44 dollars.

(B) In the general case, the problem has a solution if

$$n \geq 0 + 1 + 2 + \cdots + (p - 1) = \frac{p(p - 1)}{2}.$$

It has no solution if

$$n \leq \frac{p(p - 1)}{2} - 1 = \frac{(p + 1)(p - 2)}{2}.$$

53.2. [8; 9; 17, p. 237]

The general law is:

$$\frac{1}{2!} + \frac{2}{3!} + \cdots + \frac{n}{(n + 1)!} = 1 - \frac{1}{(n + 1)!}.$$

The step from n to $n + 1$ requires that

$$\frac{n + 1}{(n + 2)!} = -\frac{1}{(n + 2)!} + \frac{1}{(n + 1)!}$$

or

$$\frac{n + 2}{(n + 2)!} = \frac{1}{(n + 1)!},$$

which is true for $n = 1, 2, 3, \ldots$

53.3. [8; 9; 17, pp. 235–236] Observe that the first and fourth equations are related in the same way as the second and third equations: the left-hand sides have the same coefficients but in opposite order, and the right-hand sides are opposite. Adding the first equation to the fourth and the second to the third, we obtain

$$6(x + u) + 10(y + v) = 0$$
$$10(x + u) + 10(y + v) = 0,$$

which yields $x + u = 0$ and $y + v = 0$. Substituting $-x$ for u and $-y$ for v in the original equations, we find

$$-4x + 4y = 16$$
$$6x - 2y = -16.$$

Therefore $x = -2, y = 2, u = 2, v = -2$.

53.4. [8; 9; Cf. 20, p. 49, ex. 2.35]

(A) In setting up equations, we think of x and the angles α, β, γ, δ as given, and l as the unknown. From $\triangle UVG$ we find GV in terms of x, $\alpha + \beta$, and γ (law of sines). From $\triangle VUH$ we find HV in terms of x, β, and $\gamma + \delta$ (law of sines). From $\triangle GHV$ we find l in terms of GV, HV, and δ (law of cosines), and using the expressions for GV and HV, we obtain

$$l^2 = x^2 \left[\frac{\sin^2(\alpha + \beta)}{\sin^2(\alpha + \beta + \gamma)} + \frac{\sin^2 \beta}{\sin^2(\beta + \gamma + \delta)} - \frac{2 \sin(\alpha + \beta) \sin \beta \cos \delta}{\sin(\alpha + \beta + \gamma) \sin(\beta + \gamma + \delta)} \right].$$

(B) In the particular case in which $\alpha = \delta$, $\beta = \gamma$, $\alpha + \beta = \gamma + \delta = \pi/2$, the above equation yields $x = l$, as it should. One can also consider the case in which $\alpha = \delta$, $\beta = \gamma$, but the value for $\alpha + \beta$ is not prescribed, and the case in which δ, γ, β, and α are substituted for α, β, γ, and δ, respectively. Finally [see 17, pp. 202–205], one can also "test by dimension."

(C) The unknown and one of the data exchange roles.

SOLUTIONS 54

54.1. [10; 17, p. 237] In the nth line the right-hand side seems to be n^3 and the left-hand side a sum of n terms. The final term of this sum in the mth odd number, or $2m - 1$, where

$$m = \frac{n(n + 1)}{2}.$$

Hence the final term of the sum on the left-hand side should be

$$2m - 1 = n^2 + n - 1.$$

The initial term of the sum considered can be derived in two ways, first by going back $n - 1$ steps from the final term, then by advancing one step from the final term of the preceding line:

$$(n^2 + n - 1) - 2(n - 1) = n^2 - n + 1$$

$$[(n - 1)^2 + (n - 1) - 1] + 2 = n^2 - n + 1.$$

But $(n^2 - n + 1) + (n^2 - n + 3) + \cdots + (n^2 + n - 1)$ is the sum of n successive terms of an arithmetic progression whose common difference is 2. This sum is

$$\frac{(n^2 - n + 1) + (n^2 + n - 1)}{2} \cdot n = n^3.$$

(A quick verification is to observe that the nth row contains n terms of which the average or "middle" term is n^2.)

54.2. [10; 17, pp. 237–238]   The perimeter of the hexagon consists of $6n$ boundary lines of length 1 and contains $6n$ vertices. Hence

$$V = 1 + 6(1 + 2 + 3 + \cdots + n) = 3n^2 + 3n + 1.$$

By 3 diagonals through its center the hexagon is divided into 6 (large) equilateral triangles. By inspection of one of these

$$T = 6(1 + 3 + 5 + \cdots + 2n - 1) = 6n^2.$$

The T triangles have jointly $3T$ sides. In this total $3T$ each internal line of division of length 1 is counted twice, whereas the $6n$ lines along the perimeter of the hexagon are counted but once. Hence

$$2L = 3T + 6n, \qquad L = 9n^2 + 3n.$$

(It follows from Euler's theorem on polyhedra that $T + V = L + 1$.)

54.3. [10; 20, p. 54]   Expanding the right-hand side of the hypothetical identity and equating corresponding coefficients, we obtain

$$aA = bB = cC = 1 \tag{1}$$
$$bC + cB = cA + aC = aB + bA = 0. \tag{2}$$

We derive from (2) that

$$bC = -cB, \quad cA = -aC, \quad aB = -bA,$$

and multiplying these three equations, we derive further that

$$abcABC = -abcABC$$

or

$$abcABC = 0.$$

Yet we derive from (1) that

$$abcABC = 1.$$

Consequently, the hypothetical identity is impossible.

SOLUTIONS 55

55.1. [11; 17, p. 234]   The desired piece of land is bounded by two meridians and two parallel circles. Since the arc of a parallel circle intercepted by two fixed meridians is steadily shortened as the circle moves away from the equator, the center of the land Bob wants must lie on the equator; he cannot get it in the United States.

55.2. [11; 20, p. 54]

(A) Comparing coefficients of like powers on both sides of the identity, we obtain

$$1 = p^2, \qquad 4 = 2pq, \qquad -2 = q^2 + 2pr,$$
$$-12 = 2qr, \qquad 9 = r^2.$$

The first three equations, used successively, determine two systems of solutions

$$p = 1, q = 2, r = -3, \quad \text{and} \quad p = -1, q = -2, r = 3,$$

both of which also satisfy the remaining two equations.

(B) Usually it is not possible to satisfy a system with more equations than unknowns.

55.3. [11; 17, p. 236; Cf. 20, pp. 53–54, ex. 2.60–2.61]

(A) Bob, Paul, and Peter traveled equal distances, so we have

$$ct_1 - ct_2 + ct_3 = ct_1 + pt_2 + pt_3 = pt_1 + pt_2 + ct_3.$$

The second equation yields

$$(c - p)t_1 = (c - p)t_3.$$

Since we assume $c > p$, it follows that $t_1 = t_3$, that is, Peter walks just as much as Paul. From the first equation, we find that

$$(c - p)t_3 = (c + p)t_2$$

and so we obtain

$$\frac{t_1}{t_2} = \frac{t_3}{t_2} = \frac{c + p}{c - p}.$$

Hence the progress per hour is

$$\frac{c(t_1 - t_2 + t_3)}{t_1 + t_2 + t_3} = \frac{c(c + 3p)}{3c + p}.$$

(B) $\dfrac{t_2}{t_1 + t_2 + t_3} = \dfrac{c - p}{3c + p}.$

(C) In the extreme case $p = 0$, (A) yields $c/3$ and (B) yields $1/3$. If $p = c$, (A) yields c and (B) yields 0. These values are intuitively reasonable.

55.4. [11; 17, p. 235] The base of the pyramid is a polygon with n sides. In the case (A), the n lateral edges of the pyramid are equal; in the case (B), the altitudes (drawn from the apex) of its n lateral faces are equal. Draw the altitude of the pyramid, and join its foot to the n vertices of the base in case (A), but to the feet of the altitudes of the n lateral faces in case (B). In both cases, we obtain n congruent right triangles. They have one leg (the altitude of the pyramid) in common; and the hypotenuse—a lateral edge in case (A), a lateral altitude in case (B)—is of the same length in each. Consequently, the third sides in the congruent triangles must be equal. Since they are drawn from the same point (the foot of the altitude) in the same plane (the base), they form n radii of a circle which is circumscribed about, or inscribed into, the base of the pyramid, in cases (A) and (B) respectively. In case (B), it remains to show that the n radii mentioned are perpendicular to the respective sides of the base, but this follows from a well-known theorem of solid geometry.

SOLUTIONS 56

56.1. [12; 17, p. 234] Any plane figure with a center of symmetry is divided by a straight line through this center into two congruent parts, hence two parts of equal area. The required line passes through the center of symmetry.

56.2. [12; Cf. 17, p. 238, prob. 20, and 20, p. 97, ex. 3.84] Denote the number of ways to pay an amount of n cents as

A_n if only cents are used,
B_n if cents and nickels are used,
C_n if cents, nickels, and dimes are used,
D_n if cents, nickels, dimes, and quarters are used,
E_n if cents, nickels, dimes, quarters, and half-dollars are used.

(You may see now the reason for the notation E_n.) Consider the case of C_n. If no dime is used, the number of ways is B_n. If at least one dime is used, $n - 10$ cents remain to be paid in cents, nickels, and dimes. Hence

$$C_n = B_n + C_{n-10}$$

Similarly

$$B_n = A_n + B_{n-5}$$
$$D_n = C_n + D_{n-25}$$
$$E_n = D_n + E_{n-50}.$$

These formulas remain valid if we set

$$A_0 = B_0 = C_0 = D_0 = E_0 = 1$$

and regard any one of the quantities A_n, B_n, . . . , E_n as equal to 0 for n negative. The formulas allow us to compute the quantities considered *recursively*, that is, by going back to lower values of n or to former letters of the alphabet. This yields the following table which contains among others the values for E_{25} ($= D_{25}$) and E_{50}.

n	0	5	10	15	20	25	30	35	40	45	50
A_n	1	1	1	1	1	1	1	1	1	1	1
B_n	1	2	3	4	5	6	7	8	9	10	11
C_n	1	2	4	6	9	12	16	20	25	30	36
D_n	1	2	4	6	9	13	18	24	31	39	49
E_n	1	2	4	6	9	13	18	24	31	39	50

56.3. [12; Cf. 21, pp. 26–28, sect. 8.4, and p. 32, ex. 8.3] Let α denote the angle opposite side a of $\triangle$. The equal sides of the isosceles triangles with bases b and c have lengths $b/\sqrt{3}$ and $c/\sqrt{3}$, respectively. Two of these sides form with s a triangle such that the angle opposite s is $\alpha + \pi/3$. The law of cosines applied to this triangle yields

$$3s^2 = b^2 + c^2 - 2bc \cos \left(\alpha + \frac{\pi}{3}\right).$$

Apply the law of cosines to the given triangle $\triangle$ to express $bc \cos \alpha$, and set $bc \sin \alpha = 2T$, where T is the area of $\triangle$, and obtain

$$6s^2 = a^2 + b^2 + c^2 + 4\sqrt{3}\,T.$$

Since T is symmetric in a, b, and c, so is this expression.

56.4. [12] Let A, B, C, D, . . . , J be the persons around the table, and a, b, c, d, . . . , j the amounts received by them, respectively; B is to the right of A, C to the right of B, . . . , A to the right of J. The rule is expressed by the equations

$$b = \frac{a+c}{2}, \qquad c = \frac{b+d}{2}, \qquad d = \frac{c+e}{2}, \ldots, \qquad a = \frac{j+b}{2}.$$

First solution. From the above equations, it follows that

$$b - a = c - b = d - c = \ldots = a - j$$

so that everyone's share exceeds that of his neighbor on the left by the same amount. This constant excess must be zero, since

$$(b - a) + (c - b) + (d - c) + \ldots + (a - j) = 0.$$

There is just one way to distribute the money: all shares are equal. *Second solution.* Some person (or persons) must receive the maximum amount. Let such a person be B. Then none of the numbers a, . . . , j is greater than b; and, in particular,

$$b - a \geqq 0, \qquad b - c \geqq 0.$$

Yet, by the condition,

$$b - a = -(b - c).$$

Consequently, both of the two numbers $b - a$ and $b - c$ must be zero. Thus c also attains the maximum, as does d, and so on. Therefore $a = b = c = \ldots = j$.

57.1. [13; 20, p. 55] Bob has x stamps of which y sevenths are in the second book; x and y are positive integers,

$$\frac{2x}{10} + \frac{yx}{7} + 303 = x$$

and hence

$$x = \frac{3 \cdot 5 \cdot 7 \cdot 101}{28 - 5y}.$$

The denominator on the right-hand side must be positive and *odd*, since it must divide the numerator, which is odd. This leaves three possibilities: $y = 1$, 3, and 5. Only the last case yields a divisor of the numerator. Therefore the unknowns are uniquely determined: $y = 5$ and $x = 3535$.

57.2. [13; Cf. 20, p. 34, sect. 2.5 (2)] A plane passing through h and the trirectangular vertex intersects the tetrahedron in a right triangle with hypotenuse h, one leg p, and the other leg, say k, which is the altitude perpendicular to side a in the triangle whose area is A. Therefore

$$h^2 = k^2 + p^2 \qquad \text{and} \qquad A = \frac{1}{2}\, ak.$$

Since $2D = ah$, it follows from the last two equations that

$$4D^2 = a^2 h^2 = a^2(k^2 + p^2) = 4A^2 + a^2 p^2.$$

Using an equation established before (in the first plan given in the hint), we obtain further

$$4D^2 = 4A^2 + (r^2 + q^2)p^2 = 4A^2 + (rp)^2 + (pq)^2.$$

Using two more equations established before and dividing by 4, we obtain finally

$$D^2 = A^2 + B^2 + C^2,$$

which is analogous to Pythagoras' theorem.

57.3. [13; 20, p. 50] Consider the simplest special case first, that of the equilateral triangle. Symmetry may lead us to suspect that in this case the four triangular pieces will also be equilateral. If this is so, however, the sides of the triangular pieces must be *parallel* to the sides of the given triangle. This observation leads to a configuration that solves the general case as well as the particular case of the equilateral triangle: by four parallels to a side of the given triangle, dissect each of the other two sides into five equal segments.

Performing this construction three times, with respect to each side of the given triangle, we divide it into 25 congruent triangles similar to it. From these 25 triangles, we easily pick out the four mentioned in the problem; the area of each of them is 1/25 of the given triangle's area. (The uniqueness of this solution is not proved.)

SOLUTIONS 58

58.1. [14; 20, p. 139] We decompose 32118 into prime factors as $2 \times 3 \times 53 \times 101$. There are just six ways to decompose this number into a product of three factors all different from 1:

$$
\begin{array}{ll}
6 \times 53 \times 101 & 2 \times 101 \times 159 \\
3 \times 101 \times 106 & 2 \times 53 \times 303 \\
3 \times 53 \times 202 & 2 \times 3 \times 5353
\end{array}
$$

Only the first of these decompositions presents two factors between 4 and 100. Therefore, the captain has 6 children, he is 53 years old, and the length of his boat is 101 feet.

58.2. [14; Cf. 20, p. 153, ex. 6.24] Setting $x + y + u + v = s$ (which remains unchanged by any permutation of x, y, u, and v), and adding the four equations, we obtain

$$s = -3$$

and the system reduces to the following four equations:

$$s - v = 4, \quad s - x = -5, \quad s - y = 0, \quad s - u = -8.$$

Consequently, $x = 2$, $y = -3$, $u = 5$, and $v = -7$.

58.3. [14; 19, p. 214, ex. 16.6.1] Assertion I is false; assertions II and III are true.

When the three altitudes, medians, or bisectors all lie entirely within the triangle (as they do, except in the case of the altitudes in an obtuse or right triangle), we have the following situation: A, B, C are the vertices of the triangle, and A', B', C' are interior points on the opposite sides, respectively; and we have to examine the sum $AA' + BB' + CC'$. Since the sum of any two sides of a triangle is greater than the third side,

$$AA' + A'B > AB$$
$$AA' + A'C > AC.$$

Adding, we obtain

$$2AA' + BC > AB + CA$$

and by analogy

$$2BB' + CA > BC + AB$$
$$2CC' + AB > CA + BC.$$

Adding the last three inequalities, we obtain

$$2(AA' + BB' + CC') > AB + BC + CA.$$

Consequently, assertions II and III are true, and assertion I is true for acute triangles.

As a counterexample to assertion I, consider an isosceles triangle with base b and base angles A. As A tends to 0, each altitude tends to 0, whereas the perimeter tends to $2b$. An angle A sufficiently close to 0 refutes assertion I.

58.4. [14] The cases observed suggest the guess

$$1!1 + 2!2 + 3!3 + \ldots + n!n = (n+1)! - 1.$$

This can be proved by mathematical induction or in the following way:

$$(n+1)! - n! = n!(n+1) - n! = n!n.$$

Therefore, for $n = 1, 2, 3, \ldots, n$:

$$1!1 = 2! - 1!$$
$$2!2 = 3! - 2!$$
$$3!3 = 4! - 3!$$
$$\cdots$$
$$n!n = (n+1)! - n!$$

Addition of these equations yields, after obvious cancellations, the desired result.

SOLUTIONS 59

59.1. [20, p. 53] We use the following notation:

u for Al's speed,
v for Bill's speed,
t_1 for the time from the start to the first meeting,
t_2 for the time from the start to the second meeting,
d for the length of the street.

Then

$$ut_1 = a \qquad\qquad ut_2 = d + b$$
$$vt_1 = d - a \qquad\qquad vt_2 = 2d - b$$

(A) By expressing u/v in two different ways, we obtain

$$\frac{a}{d-a} = \frac{d+b}{2d-b}.$$

Hence, after discarding the vanishing root, we find $d = 3a - b$.

(B) Al walks faster. Numerically: $u/v = 3/2$.

59.2. [20, p. 50] For the first pattern $100\pi/4$, and for the second $100\pi/(2\sqrt{3})$, or approximately 78.54% and 90.69% respectively. The transition from a large (square) table to the infinite actually involves the concept of limit, but we do not insist on this since the result is intuitive.

59.3. We observe first that if n is greater than 1, the quotient of $n^{n-1} - 1$ and $n - 1$ is

$$n^{n-2} + n^{n-3} + \cdots + n + 1.$$

We conceive this sum as resulting from the polynomial

$$R(x) = (1+x)^{n-2} + (1+x)^{n-3} + \cdots + (1+x) + 1$$

when we substitute in it $n - 1$ for x. In the expansion of $R(x)$ in powers of x (you may use the binomial formula), the term independent of x is $R(0) = n - 1$, and so

$$R(x) = Q_{n-3}(x)x + n - 1$$

where $Q_{n-3}(x)$ is a polynomial of degree $n - 3$ whose coefficients are *integers*. Now we substitute $n - 1$ for x and collect our conclusions:

$$n^{n-1} - 1 = (n-1) R(n-1)$$
$$= (n-1) [Q_{n-3}(n-1)(n-1) + n - 1]$$
$$= (n-1)^2 [Q_{n-3}(n-1) + 1].$$

59.4. Let A, B, and C be the vertices of the given triangle, and let a, b, and c be the lengths of the opposite sides, respectively. Consider the side of the hexagon opposite the side of length a, and denote its length by a'. By drawing various figures and considering extreme cases, we see that we must show $a' = 2a_m$, where a_m is the length of the median of $\triangle ABC$ drawn to the side of length a. Extend AC to D so that $AD = b$, and join D to B. Now $\triangle CDB \sim \triangle CAA'$, where A' is the midpoint of the side BC so that $AA' = a_m$. Hence the median of $\triangle ABC$ of length a_m is parallel to BD and $a_m = BD/2$. But the triangle whose sides have lengths a', b, and c is congruent to $\triangle ABD$ (two sides and an included angle). Hence $a_m = a'/2$.

60.1. [15; 20, p. 55] If the reduced price is x cents and there are y pens in the remaining stock, $x < 50$ and
$$xy = 3193.$$
Now, $3193 = 31 \times 103$ is a product of two prime factors, and so it has precisely four different factors, 1, 31, 103, and 3193. If we *assume* that x is an integer, $x = 1$ or 31. If we *assume also* that $x > 1$, then $x = 31$.

60.2. [15; 20, p. 50] Let the distances of P from the four sides of the rectangle be x, y, x', y', in cyclical order. With notation appropriately chosen,
$$a^2 = y'^2 + x^2, \qquad b^2 = x^2 + y^2,$$
$$c^2 = y^2 + x'^2, \qquad d^2 = x'^2 + y'^2,$$
and so
$$a^2 + c^2 = b^2 + d^2.$$
In our case, $a = 5$, $b = 10$, $c = 14$, and so
$$d^2 = 25 - 100 + 196 = 121, \qquad d = 11.$$
Observe that the data a, b, and c which determine d are insufficient to determine the sides $x + x'$ and $y + y'$ of the rectangle.

60.3. [15] We write the known relation
$$\sin 2\alpha = 2 \sin \alpha \cos \alpha$$
and substitute in it successively $\alpha/2$, $\alpha/4$, and $\alpha/8$ for α. We obtain
$$\cos \frac{\alpha}{2} = \frac{\sin \alpha}{2 \sin \dfrac{\alpha}{2}}$$

$$\cos \frac{\alpha}{4} = \frac{\sin \dfrac{\alpha}{2}}{2 \sin \dfrac{\alpha}{4}}$$

$$\cos \frac{\alpha}{8} = \frac{\sin \dfrac{\alpha}{4}}{2 \sin \dfrac{\alpha}{8}}.$$

Multiplication of the three equations yields, after obvious cancellations, the required identity. If we had proceeded to n successive equations instead of three, the multiplication would have yielded
$$\cos \frac{\alpha}{2} \cos \frac{\alpha}{4} \cos \frac{\alpha}{8} \cdots \cos \frac{\alpha}{2^n} = \frac{\sin \alpha}{2^n \sin \dfrac{\alpha}{2^n}}.$$

54

One can also guess this more general formula, and having guessed it, prove it afterwards by mathematical induction.

60.4. [15; 20, p. 51] Let O stand for the volume of the octahedron, T for the volume of the tetrahedron, and a for the length of an edge.

First solution. The octahedron is divided by an appropriate plane into two congruent regular pyramids whose common square base has area a^2. The height of one of these pyramids is $a/\sqrt{2}$ (the "key plane figure" passes through a diagonal of the base) and so

$$O = 2\,\frac{a^2}{3}\,\frac{a}{\sqrt{2}} = \frac{a^3\sqrt{2}}{3}.$$

Pass a plane through the altitude (of length h) of the tetrahedron and through a coterminal edge. The intersection (the key plane figure) is divided by the altitude into two right triangles, from which we obtain

$$h^2 = a^2 - \left(\frac{2a\sqrt{3}}{6}\right)^2 = \left(\frac{a\sqrt{3}}{2}\right)^2 - \left(\frac{a\sqrt{3}}{6}\right)^2 = \frac{2a^2}{3}$$

and so

$$T = \frac{1}{3}\,\frac{a}{2}\,\frac{a\sqrt{3}}{2}\,\frac{a\sqrt{2}}{\sqrt{3}} = \frac{a^3\sqrt{2}}{12}.$$

Finally,

$$O = 4T.$$

Second solution. Consider the regular tetrahedron with edge $2a$; its volume is $2^3 T$. Four planes, each of which passes through the midpoints of three of its edges terminating in the same vertex, dissect it into four regular tetrahedra, each of volume T, and a regular tetrahedron of volume O. Hence,

$$4T + O = 8T$$

which yields again $O = 4T$.

SOLUTIONS 61

61.1. [16; Cf. 20, p. 153, ex. 6.24] The simplest expression that is symmetric in x, y, and z is their sum. Adding the three proposed equations, we obtain

$$10000x + 10000y + 10000z = 20000,$$
$$x + y + z = 2.$$

By subtracting

$$2134x + 2134y + 2134z = 4268$$

from each of the three proposed equations, we obtain three new equations that when solved yield $x = 1$, $y = -1$, $z = 2$, respectively.

61.2. [16; 20, p. 149] The condition is easily split into two parts expressed by the two equations

$$b + g + w + s = 14$$
$$b + 2g + 3w + 4s = 30.$$

Subtracting the first from the second, we obtain

$$g + 2w + 3s = 16,$$

which shows that either g and s are both odd or they are both even. Hence there are only four cases that need to be examined:

g	s	$w = 8 - (g + 3s)/2$
3	5	-1
5	3	1
2	4	1
4	2	3

Only the last case is admissible. Therefore,

$$s = 2, \qquad w = 3, \qquad g = 4, \qquad b = 5$$

and the ladies are

Ann Smith,   Betty White,   Carol Green,   Dorothy Brown.

61.3. [16; 21, p. 193] If $x = 0$, the second (or third) equation yields $y^2z^2 = 0$, and so one more unknown, y or z, must also be 0. Hence either x, y, and z are all different from 0 or at least two vanish. If any two vanish, the equations are satisfied.

Now we consider the case in which no one of the three unknowns is 0. By dividing, we obtain the system

$$\frac{zx}{y} + \frac{xy}{z} = a,$$
$$\frac{yz}{x} \qquad\; + \frac{xy}{z} = b,$$
$$\frac{yz}{x} + \frac{zx}{y} \qquad\; = c.$$

Adding these three equations and dividing by 2, we have

$$\frac{yz}{x} + \frac{zx}{y} + \frac{xy}{z} = \frac{a + b + c}{2}.$$

From this equation we subtract each of the three equations of the foregoing system and obtain

$$\frac{yz}{x} = \frac{-a+b+c}{2}$$

$$\frac{zx}{y} = \frac{a-b+c}{2}$$

$$\frac{xy}{z} = \frac{a+b-c}{2}.$$

The product of these three equations is

$$xyz = (-a+b+c)(a-b+c)(a+b-c)/8 \qquad (*)$$

which we divide by each equation of the foregoing system to obtain, after extracting a square root,

$$x = [(a-b+c)(a+b-c)]^{\frac{1}{2}}/2$$
$$y = [(-a+b+c)(a+b-c)]^{\frac{1}{2}}/2$$
$$z = [(-a+b+c)(a-b+c)]^{\frac{1}{2}}/2.$$

We must take into account, however, the two values of each square root. Let us concentrate upon a suggestive particular case and assume that a, b, and c are the lengths of the three sides of a triangle. Then by ($*$) above, xyz is positive, and therefore only the following four combinations of signs are admissible:

x	$+$	$+$	$-$	$-$
y	$+$	$-$	$+$	$-$
z	$+$	$-$	$-$	$+$

61.4. [16; Cf. 20, p. 51, ex. 2.49] Pass a plane through the altitude of the pyramid and through the midpoint of one side (of length a) of its base. The intersection of this plane with the pyramid is an isosceles triangle that can be used as a key figure: its height is h, its legs are of length l (where l is the height of a lateral face of the pyramid), and its base is $2b$ (where b is the altitude of one of the six congruent equilateral triangles composing the base of the pyramid). The area of the base is

$$\frac{S}{4} = \frac{6ab}{2},$$

the area of the lateral surface is

$$\frac{3S}{4} = \frac{6al}{2},$$

and so

$$l = 3b.$$

Using the key figure, we obtain
$$h^2 + b^2 = l^2 = 9b^2$$
and so
$$b^2 = \frac{h^2}{8}.$$

We also have
$$b^2 + \frac{a^2}{4} = a^2$$
and so
$$a^2 = \frac{4b^2}{3} = \frac{h^2}{6}.$$

Therefore
$$S = 12ab = h^2\sqrt{3}.$$

SOLUTIONS 62

62.1. [Cf. 21, p. 162, ex. 15.36] We are required to find the points of intersection of two congruent ellipses symmetrical to each other with respect to the line $x = y$. Subtraction of the equations yields $x^2 = y^2$. There are four points of intersection: $(6, 6)$, $(-6, -6)$, $(2, -2)$, $(-2, 2)$.

62.2. The product of the four products is symmetric in a, b, c, and d. If we can show that
$$4a(1 - b)\ 4b(1 - c)\ 4c(1 - d)\ 4d(1 - a) \leqq 1,$$
it will follow that not all of the four given products are greater than one. We are given that $0 < a < 1$, $0 < b < 1$, $0 < c < 1$, and $0 < d < 1$. Consider the product $4a(1 - a)$, which is positive as its factors are positive; what is its maximum? An obvious guess is that it is 1, attained when $a = \frac{1}{2}$. To verify this, observe that
$$1 - 4a(1 - a) = (1 - 2a)^2 \geqq 0$$
and that the case of equality is attained only when $a = \frac{1}{2}$. Similarly, the products $4b(1 - b)$, $4c(1 - c)$, and $4d(1 - d)$ are also positive and not greater than one.

Consequently,
$$4a(1 - a)\ 4b(1 - b)\ 4c(1 - c)\ 4d(1 - d) \leqq 1.$$
Rearranging terms,
$$4a(1 - b)\ 4b(1 - c)\ 4c(1 - d)\ 4d(1 - a) \leqq 1.$$

62.3.

(A) Since $\angle B'CA = \angle A'CB = 45°$ and $\angle C = 90°$, C lies on the line segment $A'B'$. $\triangle ABC$ and $\triangle ABC'$ are both right triangles; they can be inscribed in a circle with diameter AB. $\angle ACC'$ and $\angle BCC'$ intercept equal arcs of the circle, so they are equal. Therefore, $\angle B'CC' = \angle A'CC' = 90°$.

(B) If the sides of the triangle are of length a, b, and c, then $B'C$ is of length $b/\sqrt{2}$, $A'C$ is of length $a/\sqrt{2}$, and $A'B'$ is therefore of length $(a + b)/\sqrt{2}$.

Let D be the vertex opposite A in the square on side b. Then $\triangle ACC' \sim \triangle ADB$ (pairs of corresponding angles are equal). Therefore, $CC' : AC = DB : AD$, and CC' is also of length $(a + b)/\sqrt{2}$.

(A proof, using transformations, of the general case in which the triangle is not necessarily a right triangle is given in [22, pp. 96–97].)

62.4. [21, p. 188]

(A) Pass a plane through the edge of length b and the midpoint M of the opposite edge. The intersection of this plane with the tetrahedron is an isosceles triangle that can be used as a key figure: its base is of length b, its legs are of length $a\sqrt{3}/2$ (they are altitudes of the equilateral faces of the tetrahedron), and therefore its height is $\sqrt{3a^2 - b^2}/2$. By symmetry, the center O of the circumscribed sphere lies on the line joining M to the midpoint B of the base of this triangle. Also, the line joining O to the center C of one of the equilateral faces of the tetrahedron is perpendicular to that face. The point C divides each altitude (or median) of the equilateral face in the ratio $1 : 2$. Therefore CM is of length $a\sqrt{3}/6$. Triangle OCM is similar to each triangle into which the key figure is divided by MB. Therefore the ratios of corresponding sides are equal. If x is the length of OB, we have

$$\frac{\dfrac{\sqrt{3a^2 - b^2}}{2} - x}{\dfrac{a\sqrt{3}}{6}} = \frac{\dfrac{a\sqrt{3}}{2}}{\dfrac{\sqrt{3a^2 - b^2}}{2}}.$$

Solving this equation for x, we obtain

$$x = \frac{2a^2 - b^2}{2\sqrt{3a^2 - b^2}}.$$

If r is the radius of the circumscribed sphere,

$$r^2 = x^2 + \left(\frac{b}{2}\right)^2 = \frac{(2a^2 - b^2)^2}{4(3a^2 - b^2)} + \frac{b^2}{4} = \frac{a^2(4a^2 - b^2)}{4(3a^2 - b^2)}$$

and

$$r = \frac{a}{2}\sqrt{\frac{4a^2 - b^2}{3a^2 - b^2}}.$$

Observe that the denominator vanishes when $b = a\sqrt{3}$. This is a limiting case, since for the tetrahedron considered in the problem, $b < a\sqrt{3}$.

(B) The radius of a spherical surface can be determined by means of a device in which four points are arranged to form the vertices of two congruent equilateral triangles sharing a common side and "hinged" along that side. Let the sides of the triangle have length a. By placing the four points in contact with the surface and by measuring the distance b between the two points that are not endpoints of the common side, one can use the result (A) to calculate the radius of the sphere determined by the four points.

SOLUTIONS 63

63.1. [21, p. 188] Each vertex of the triangle is equidistant from two of the three points in which the sides of the triangle are tangent to the inscribed circle. Label A, B, and C as the vertices opposite sides a, b, and c, respectively. Then the distance from C to the two nearest points of tangency is $d/2$, the distance from B to the two nearest points of tangency is $a - (d/2)$, and the distance from A to the two nearest points of tangency is $b - (d/2)$. Since

$$a - (d/2) + b - (d/2) = c,$$

it follows that

$$a + b = c + d.$$

63.2. [Cf. 21, p. 191, ex. 3.65.1]

First solution. Rewrite the expression as

$$n[(n - 2)(n - 1)n(n + 1)(n + 2)]$$

and observe that the expression in brackets is the product of five consecutive integers. We observe that $360 = 2^3 \cdot 3^2 \cdot 5$. Given any five consecutive integers, one must be a multiple of 5; hence 5 divides the expression. Also, one of these five integers must be a multiple of 4, and since at least one other must be even, 8 must divide the expression. If n is a multiple of 3, then n^2 is a multiple of 9, and 9 divides the expression. If n is not a multiple of 3, then either $n - 2$ and $n + 1$ or $n - 1$ and $n + 2$ are each divisible by 3, and hence 9 divides the expression. Since the expression is divisible by 5, 8, and 9, and since these have no factors in common, the expression must be divisible by their product.

Second solution.

$$\frac{n^2(n^2 - 1)(n^2 - 4)}{360}$$

$$= \frac{[(n + 3) + (n - 3)](n + 2)(n + 1)n(n - 1)(n - 2)}{6!}$$

$$= \binom{n + 3}{6} + \binom{n + 2}{6}$$

and binomial coefficients are integers.

63.3. [Cf. 21, p. 163, ex. 15.37–15.38]

Subtracting, in turn, the second equation from the first, the third from the second, and the first from the third, we obtain the new system

$$-5x^2 + 4y^2 + z^2 = 0$$
$$x^2 - 5y^2 + 4z^2 = 0$$
$$4x^2 + y^2 - 5z^2 = 0.$$

We can eliminate z^2 by multiplying the first equation in this system by -4 and adding it to the second. We can eliminate x^2 by multiplying the second equation by -4 and adding it to the third. This yields

$$x^2 = y^2 = z^2.$$

Substitution in the original system yields eight solutions:

$$(1, 1, 1), \qquad (-1, -1, -1),$$
$$(3, -3, -3), \qquad (-3, 3, 3), \qquad (-3, 3, -3),$$
$$(3, -3, 3), \qquad (-3, -3, 3), \qquad (3, 3, -3).$$

63.4. [Cf. 21, p. 162, ex. 15.29–15.30] Let r be the radius of the circle inscribed in the base of the prism. The base is composed of six equilateral triangles whose sides are of length $2r/\sqrt{3}$. The volume of the prism is

$$6 \frac{r^2}{\sqrt{3}} 2r = 4\sqrt{3} \cdot r^3.$$

The surface area of the prism is

$$2 \cdot 6 \frac{r^2}{\sqrt{3}} + 6 \frac{2r}{\sqrt{3}} 2r = 4\sqrt{3} \cdot r^2 + 8\sqrt{3} \cdot r^2 = 12\sqrt{3} \cdot r^2.$$

Let a be the length of an edge of the regular octahedron whose volume is $4\sqrt{3} \cdot r^3$. The octahedron can be divided into two congruent pyramids whose common base is a square with side of length a and whose height is half the diagonal of a square with side of length a. Thus the volume of the octahedron is

$$2 \cdot \frac{1}{3} a^2 \frac{a}{\sqrt{2}} = \frac{a^3\sqrt{2}}{3} = 4\sqrt{3} \cdot r^3$$

and

$$a = \sqrt{6} \cdot \text{r}.$$

The surface area of the octahedron is

$$8 \frac{\sqrt{3} \cdot a}{2} \frac{a}{2} = 2\sqrt{3} \cdot a^2 = 2\sqrt{3} \cdot 6r^2 = 12\sqrt{3} \cdot r^2.$$

Therefore the two surface areas are equal.

Given two solids with the same number of faces and the same volume, one would naturally expect that if one of them is regular, it will have a smaller surface area than the other.

SOLUTIONS 64

64.1. [21, p. 188] Let s be the length of the side of the base of the given prism (the cake). Then the altitude of the given prism is $5s/16$, the volume is $5s^3/16$, and the surface (5 faces) with icing is $9s^2/4$. For the required prism (the piece of cake), the area of the base is $s^2/4$ and the volume is $5s^3/(16 \times 9)$. The altitude is $5s/36$, the quotient of the volume and the base area. The side of the base of the required prism is $s/2$. Therefore the required ratio is

$$\frac{5s}{36} \cdot \frac{2}{s} = \frac{5}{18}.$$

On the top of the cake (a square) we mark a concentric square. The sides of the smaller square are parallel to and half the length of the sides of the larger square. Line segments join corresponding vertices and the midpoints of corresponding sides.

Each of the 8 pieces with icing on the side is a right prism of which a smaller right prism is cut off; for both prisms, larger and smaller, the base is an isosceles right triangle.

Another solution can be obtained by rotating the square piece about its center through $45°$.

64.2.

First solution. Recall the formula for the sum of a geometric progression. The number of the sequence that has $2n$ digits is

$$9 + 8(10 + 10^2 + \cdots + 10^{n-1})$$
$$+ 4(10^n + 10^{n+1} + \cdots + 10^{2n-1})$$
$$= 1 + (8 + 4 \cdot 10^n)(1 + 10 + \cdots + 10^{n-1})$$
$$= 1 + 4(10^n + 2)(10^n - 1)/(10 - 1)$$
$$= \left(\frac{2 \cdot 10^n + 1}{3}\right)^2.$$

This is the square of an integer, since

$$(2 \cdot 10^n + 1)/3 = 1 + 6(10^n - 1)/9 = 666 \cdots 67,$$

a number with n digits.

Second solution. Experimentation with some examples,

$$49 = 7^2, \qquad 4489 = 67^2, \qquad 444889 = 667^2$$

leads to the conjecture that the nth term has the form $(666 \cdots 67)^2$, where $666 \cdots 67$ has n digits. To confirm the conjecture by a proof, it is possible to start from the remark that

$$666 \cdots 667 \times 6 = 4000 \cdots 002$$
$$666 \cdots 667 \times 7 = 4666 \cdots 669$$

and visualize the usual pattern of multiplication for integers written in decimal notation.

64.3. Here are easy examples, all right triangles:

	SIDES		AREA
3	4	5	6
1	1	$\sqrt{2}$	½
$\sqrt{2}$	$\sqrt{2}$	2	1
$3\sqrt{2}$	$4\sqrt{2}$	$5\sqrt{2}$	12

There are many other examples, but triangles that are not right triangles are less convenient.

64.4.

First solution. Essentially, we are considering here three sets A, B, and C, and their intersections. If each set is represented by points inside a circle (Venn diagram, or more correctly Eulerian circle), the three circles, each partly overlapping the other two, cut the plane into eight regions. One of these regions, outside all three circles, is infinite; the problem is to find the number of individuals belonging to this infinite region. Beginning with the region inside all three circles (from the right-hand end of the row of given numbers) and working outward (to the left in the row), one can calculate how many individuals are in each region. Let $\overline{X}$ denote the set of individuals not in set X, let XY denote the intersection of sets X and Y, and let $[X]$ denote the number of individuals in set X. Then the number of individuals in the union of A, B, and C is given by

$$[ABC] + [\overline{A}BC] + [A\overline{B}C] + [AB\overline{C}] + [\overline{A}\overline{B}C] + [\overline{A}B\overline{C}] + [A\overline{B}\overline{C}]$$
$$= 1 + 2 + 5 + 1 + 0 + 1 + 5 = 15$$

Thus the required number (of students passing in all three subjects) is $41 - 15 = 26$.

Second solution. Let us use the same notation as in the first solution. The number of individuals in the union of A, B, and C is also given by

$$[A] + \quad [B] \quad + [C]$$
$$- [AB] - \quad [AC] \quad - [BC]$$
$$+ [ABC]$$

An individual who belongs to exactly one of the three sets is counted once in the first line and nowhere else. An individual who belongs to exactly two of the sets is counted twice positively in the first line and once negatively in the second line. An individual who belongs to all three sets is counted three times positively in the first line, three times negatively in the second, and once positively in the third. Thus, the proposed expression counts all the individuals it must count, each just once (and it can be generalized to a larger number of sets).

Applying the general expression to our present case, we find that

$$12 + 5 + 8$$
$$-2 - 6 - 3$$
$$+ 1 \quad = 15$$

students failed in one or more subjects, and so $41 - 15 = 26$ passed in all three.

(*Note:* The problem on the original examination paper contained an error: 4 was inadvertently substituted for 2. Although the data were therefore inconsistent, contestants were nonetheless able to use methods essentially equivalent to those given above, and their arguments were judged accordingly.)

64.5. [21, p. 195]

(A) Our task is greatly facilitated if we have learned and remember a classical proposition of geometry (Euclid VI 3): The segments of the base, p and q, have the same proportion as the adjacent sides, a and b:

$$\frac{p}{q} = \frac{a}{b}.$$

Since $p + q = c$, we have

$$p = \frac{ac}{a+b}, \qquad q = \frac{bc}{a+b}.$$

Applying the law of cosines twice to two different triangles that both contain the angle α, opposite to the side a in the original triangle:

$$a^2 = b^2 + c^2 - 2bc \cos \alpha$$
$$d^2 = b^2 + q^2 - 2bq \cos \alpha$$

Eliminating cos α and solving for d^2, we obtain

$$d^2 = ab \left[1 - \left(\frac{c}{a+b} \right)^2 \right].$$

(If you do not know, or cannot remember, that proposition about the ratio p/q, it can be discovered by surveying trigonometric relations in the three triangles. It follows, in fact, from the law of sines applied twice to the two smaller triangles.)

(B) *A particular case.* If $a = b$, the given triangle is isosceles with base c, and the formula yields

$$d^2 = a^2 - \frac{c^2}{4}.$$

A limiting case. If the given triangle degenerates, collapses into the line segment c, then $a + b = c$ and the formula yields $d = 0$.

The formula can also be "tested by dimension" [17, pp. 202–205].

SOLUTIONS 65

65.1. [21, p. 192] If you list all twelve decompositions of 72 into three factors and note the sum of the factors, you will observe that the only sum occurring twice is 14 ($= 2\cdot6\cdot6 = 3\cdot3\cdot8$). Hence the street number is 14. In view of *the* oldest boy, the ages are 3, 3, and 8.

65.2. [21, p. 187] Let a and b denote the lengths of the legs. The hexagon consists of three squares, of area a^2, b^2, and c^2, respectively, and four triangles all of the same area A. Either introduce auxiliary line segments into the figure to prove that one of the two obtuse triangles has altitude a and base b and the other has altitude b and base a, or use trigonometry:

$$A = \frac{1}{2}\,ab = \frac{1}{2}\,ac\sin\beta = \frac{1}{2}\,ac\sin(180° - \beta).$$

Hence the area of the hexagon is

$$a^2 + b^2 + c^2 + 4A = 2c^2 + 4A.$$

65.3. [Cf. 21, pp. 146–149, and p. 158, ex. 15.3]

(A) Triangle with vertices $(1, 1)$, $(0, 1)$, $(½, ½)$ and sides located on the straight lines $y = 1$, $x = y$; yet $x + y > 1$

(B) Two sides of triangle (A) on lines $y = 1$ and $x = y$

(C) Arc of circle $x^2 + y^2 = 1$ inside triangle (A)

(D) Part of triangle (A) above arc (C)

(E) Part of triangle (A) below arc (C).

The point $(1, 1)$ represents the equilateral triangle; the point $(1/\sqrt{2},\, 1/\sqrt{2})$ represents the isosceles right triangle; the point $(1/2,\, \sqrt{3}/2)$ represents the 30°60°90° triangle; and the side of triangle (A) on line $x + y = 1$ represents the degenerate triangles.

65.4. [Cf. 21, p. 139, ex. 14.24] The remainder $r(x)$ is a polynomial of degree not exceeding 2:

$$r(x) = a + bx + cx^2.$$

Then

$$x + x^9 + x^{25} + x^{49} + x^{81} = q(x)\,(x^3 - x) + a + bx + cx^2.$$

This yields for $x = -1, 0$, and 1 respectively:

$$-5 = a - b + c$$
$$0 = a$$
$$5 = a + b + c.$$

Therefore $a = 0$, $b = 5$, $c = 0$, and the desired remainder is $5x$.

REFERENCES

1. R. Creighton Buck, A look at mathematical competitions, *American Mathematical Monthly*, 66 (1959) 201–212.

2. Department of Mathematics, Stanford University, The Stanford University Mathematics Examination, *American Mathematical Monthly*, 53 (1946) 406–409.

3. ———, Stanford University Mathematics Examination, *American Mathematical Monthly*, 54 (1947) 430.

4. ———, Stanford University Competitive Examination, *American Mathematical Monthly*, 55 (1948) 448.

5. ———, Stanford University Competitive Examination in Mathematics, *American Mathematical Monthly*, 56 (1949) 496–497.

6. ———, Stanford University Competitive Examination in Mathematics, *American Mathematical Monthly*, 57 (1950) 651–652.

7. ———, Stanford University Competitive Examination in Mathematics, *American Mathematical Monthly*, 59 (1952) 127.

8. ———, Stanford University Competitive Examination in Mathematics, *American Mathematical Monthly*, 60 (1953) 571–572.

9. G. Polya, The 1953 Stanford Competitive Examination: problems, solutions, and comments, *California Mathematics Council Bulletin*, (1) 11 (1953) 15–17.

10. ———, The 1954 Stanford Competitive Examination: problems and solutions, *California Mathematics Council Bulletin*, (2) 12 (1954) 7–8.

11. ———, The 1955 Stanford Competitive Examination in Mathematics: problems, solutions, and comments, *California Mathematics Council Bulletin*, (2) 13 (1955) 15–17.

12. ———, The 1956 Stanford Competitive Examination in Mathematics: solutions and comments, *California Mathematics Council Bulletin*, (2) 14 (1956) 19–22.

13. ———, The 1957 Stanford University Competitive Examination in Mathematics, *California Mathematics Council Bulletin*, (2) 15 (1957) 18–21.

14. ———, The 1958 Stanford-Sylvania Competitive Examination in Mathematics, *California Mathematics Council Bulletin*, (1) 16 (1958) 18–20.

15. ———, The 1960 Stanford-Sylvania Competitive Examination in Mathematics, *California Mathematics Council Bulletin*, (2) 18 (1960) 16–17.

68

16. ———, The 1961 Stanford-Sylvania Competitive Examination in Mathematics, *California Mathematics Council Bulletin,* (2) 19 (1961) 10–11.

17. ———, *How to Solve It,* 2nd edition, Doubleday Anchor A 93, 1957, Princeton Paperback, 1971.

18. ———, *Mathematics and Plausible Reasoning,* Vol. 1, Princeton Univ. Press, 1954.

19. ———, *Mathematics and Plausible Reasoning,* Vol. 2, 2nd edition, Princeton Univ. Press, 1968.

20. ———, *Mathematical Discovery,* Vol. 1, Wiley, 1962.

21. ———, *Mathematical Discovery,* Vol. 2, corrected printing, Wiley, 1968.

22. H. S. M. Coxeter and S. L. Greitzer, *Geometry Revisited,* Random House, 1967.

23. *Hungarian Problem Book I, II,* Random House, 1963.

24. G. Polya and J. Kilpatrick, The Stanford University Competitive Examination in Mathematics, *American Mathematical Monthly,* 80 (1973) 627–640.

A CATALOG OF SELECTED
DOVER BOOKS
IN SCIENCE AND MATHEMATICS

Mathematics

FUNCTIONAL ANALYSIS (Second Corrected Edition), George Bachman and Lawrence Narici. Excellent treatment of subject geared toward students with background in linear algebra, advanced calculus, physics and engineering. Text covers introduction to inner-product spaces, normed, metric spaces, and topological spaces; complete orthonormal sets, the Hahn-Banach Theorem and its consequences, and many other related subjects. 1966 ed. 544pp. 6⅛ x 9¼. 0-486-40251-7

ASYMPTOTIC EXPANSIONS OF INTEGRALS, Norman Bleistein & Richard A. Handelsman. Best introduction to important field with applications in a variety of scientific disciplines. New preface. Problems. Diagrams. Tables. Bibliography. Index. 448pp. 5⅜ x 8½. 0-486-65082-0

VECTOR AND TENSOR ANALYSIS WITH APPLICATIONS, A. I. Borisenko and I. E. Tarapov. Concise introduction. Worked-out problems, solutions, exercises. 257pp. 5⅜ x 8¼. 0-486-63833-2

AN INTRODUCTION TO ORDINARY DIFFERENTIAL EQUATIONS, Earl A. Coddington. A thorough and systematic first course in elementary differential equations for undergraduates in mathematics and science, with many exercises and problems (with answers). Index. 304pp. 5⅜ x 8½. 0-486-65942-9

FOURIER SERIES AND ORTHOGONAL FUNCTIONS, Harry F. Davis. An incisive text combining theory and practical example to introduce Fourier series, orthogonal functions and applications of the Fourier method to boundary-value problems. 570 exercises. Answers and notes. 416pp. 5⅜ x 8½. 0-486-65973-9

COMPUTABILITY AND UNSOLVABILITY, Martin Davis. Classic graduate-level introduction to theory of computability, usually referred to as theory of recurrent functions. New preface and appendix. 288pp. 5⅜ x 8½. 0-486-61471-9

ASYMPTOTIC METHODS IN ANALYSIS, N. G. de Bruijn. An inexpensive, comprehensive guide to asymptotic methods—the pioneering work that teaches by explaining worked examples in detail. Index. 224pp. 5⅜ x 8½ 0-486-64221-6

APPLIED COMPLEX VARIABLES, John W. Dettman. Step-by-step coverage of fundamentals of analytic function theory—plus lucid exposition of five important applications: Potential Theory; Ordinary Differential Equations; Fourier Transforms; Laplace Transforms; Asymptotic Expansions. 66 figures. Exercises at chapter ends. 512pp. 5⅜ x 8½. 0-486-64670-X

INTRODUCTION TO LINEAR ALGEBRA AND DIFFERENTIAL EQUATIONS, John W. Dettman. Excellent text covers complex numbers, determinants, orthonormal bases, Laplace transforms, much more. Exercises with solutions. Undergraduate level. 416pp. 5⅜ x 8½. 0-486-65191-6

RIEMANN'S ZETA FUNCTION, H. M. Edwards. Superb, high-level study of landmark 1859 publication entitled "On the Number of Primes Less Than a Given Magnitude" traces developments in mathematical theory that it inspired. xiv+315pp. 5⅜ x 8½. 0-486-41740-9

CALCULUS OF VARIATIONS WITH APPLICATIONS, George M. Ewing. Applications-oriented introduction to variational theory develops insight and promotes understanding of specialized books, research papers. Suitable for advanced undergraduate/graduate students as primary, supplementary text. 352pp. 5⅜ x 8½.
0-486-64856-7

COMPLEX VARIABLES, Francis J. Flanigan. Unusual approach, delaying complex algebra till harmonic functions have been analyzed from real variable viewpoint. Includes problems with answers. 364pp. 5⅜ x 8½.
0-486-61388-7

AN INTRODUCTION TO THE CALCULUS OF VARIATIONS, Charles Fox. Graduate-level text covers variations of an integral, isoperimetrical problems, least action, special relativity, approximations, more. References. 279pp. 5⅜ x 8½.
0-486-65499-0

COUNTEREXAMPLES IN ANALYSIS, Bernard R. Gelbaum and John M. H. Olmsted. These counterexamples deal mostly with the part of analysis known as "real variables." The first half covers the real number system, and the second half encompasses higher dimensions. 1962 edition. xxiv+198pp. 5⅜ x 8½. 0-486-42875-3

CATASTROPHE THEORY FOR SCIENTISTS AND ENGINEERS, Robert Gilmore. Advanced-level treatment describes mathematics of theory grounded in the work of Poincaré, R. Thom, other mathematicians. Also important applications to problems in mathematics, physics, chemistry and engineering. 1981 edition. References. 28 tables. 397 black-and-white illustrations. xvii + 666pp. 6⅛ x 9¼.
0-486-67539-4

INTRODUCTION TO DIFFERENCE EQUATIONS, Samuel Goldberg. Exceptionally clear exposition of important discipline with applications to sociology, psychology, economics. Many illustrative examples; over 250 problems. 260pp. 5⅜ x 8½.
0-486-65084-7

NUMERICAL METHODS FOR SCIENTISTS AND ENGINEERS, Richard Hamming. Classic text stresses frequency approach in coverage of algorithms, polynomial approximation, Fourier approximation, exponential approximation, other topics. Revised and enlarged 2nd edition. 721pp. 5⅜ x 8½. 0-486-65241-6

INTRODUCTION TO NUMERICAL ANALYSIS (2nd Edition), F. B. Hildebrand. Classic, fundamental treatment covers computation, approximation, interpolation, numerical differentiation and integration, other topics. 150 new problems. 669pp. 5⅜ x 8½.
0-486-65363-3

THREE PEARLS OF NUMBER THEORY, A. Y. Khinchin. Three compelling puzzles require proof of a basic law governing the world of numbers. Challenges concern van der Waerden's theorem, the Landau-Schnirelmann hypothesis and Mann's theorem, and a solution to Waring's problem. Solutions included. 64pp. 5⅜ x 8½.
0-486-40026-3

THE PHILOSOPHY OF MATHEMATICS: AN INTRODUCTORY ESSAY, Stephan Körner. Surveys the views of Plato, Aristotle, Leibniz & Kant concerning propositions and theories of applied and pure mathematics. Introduction. Two appendices. Index. 198pp. 5⅜ x 8½.
0-486-25048-2

INTRODUCTORY REAL ANALYSIS, A.N. Kolmogorov, S. V. Fomin. Translated by Richard A. Silverman. Self-contained, evenly paced introduction to real and functional analysis. Some 350 problems. 403pp. 5⅜ x 8½.   0-486-61226-0

APPLIED ANALYSIS, Cornelius Lanczos. Classic work on analysis and design of finite processes for approximating solution of analytical problems. Algebraic equations, matrices, harmonic analysis, quadrature methods, much more. 559pp. 5⅜ x 8½.
0-486-65656-X

AN INTRODUCTION TO ALGEBRAIC STRUCTURES, Joseph Landin. Superb self-contained text covers "abstract algebra": sets and numbers, theory of groups, theory of rings, much more. Numerous well-chosen examples, exercises. 247pp. 5⅜ x 8½.
0-486-65940-2

QUALITATIVE THEORY OF DIFFERENTIAL EQUATIONS, V. V. Nemytskii and V.V. Stepanov. Classic graduate-level text by two prominent Soviet mathematicians covers classical differential equations as well as topological dynamics and ergodic theory. Bibliographies. 523pp. 5⅜ x 8½.   0-486-65954-2

THEORY OF MATRICES, Sam Perlis. Outstanding text covering rank, nonsingularity and inverses in connection with the development of canonical matrices under the relation of equivalence, and without the intervention of determinants. Includes exercises. 237pp. 5⅜ x 8½.   0-486-66810-X

INTRODUCTION TO ANALYSIS, Maxwell Rosenlicht. Unusually clear, accessible coverage of set theory, real number system, metric spaces, continuous functions, Riemann integration, multiple integrals, more. Wide range of problems. Undergraduate level. Bibliography. 254pp. 5⅜ x 8½.   0-486-65038-3

MODERN NONLINEAR EQUATIONS, Thomas L. Saaty. Emphasizes practical solution of problems; covers seven types of equations. ". . . a welcome contribution to the existing literature...."—*Math Reviews*. 490pp. 5⅜ x 8½.   0-486-64232-1

MATRICES AND LINEAR ALGEBRA, Hans Schneider and George Phillip Barker. Basic textbook covers theory of matrices and its applications to systems of linear equations and related topics such as determinants, eigenvalues and differential equations. Numerous exercises. 432pp. 5⅜ x 8½.   0-486-66014-1

LINEAR ALGEBRA, Georgi E. Shilov. Determinants, linear spaces, matrix algebras, similar topics. For advanced undergraduates, graduates. Silverman translation. 387pp. 5⅜ x 8½.   0-486-63518-X

ELEMENTS OF REAL ANALYSIS, David A. Sprecher. Classic text covers fundamental concepts, real number system, point sets, functions of a real variable, Fourier series, much more. Over 500 exercises. 352pp. 5⅜ x 8½.   0-486-65385-4

SET THEORY AND LOGIC, Robert R. Stoll. Lucid introduction to unified theory of mathematical concepts. Set theory and logic seen as tools for conceptual understanding of real number system. 496pp. 5⅜ x 8¼.   0-486-63829-4

STATISTICAL PHYSICS, Gregory H. Wannier. Classic text combines thermodynamics, statistical mechanics and kinetic theory in one unified presentation of thermal physics. Problems with solutions. Bibliography. 532pp. 5⅜ x 8½.   0-486-65401-X

TENSOR CALCULUS, J.L. Synge and A. Schild. Widely used introductory text covers spaces and tensors, basic operations in Riemannian space, non-Riemannian spaces, etc. 324pp. 5⅜ x 8¼.   0-486-63612-7

ORDINARY DIFFERENTIAL EQUATIONS, Morris Tenenbaum and Harry Pollard. Exhaustive survey of ordinary differential equations for undergraduates in mathematics, engineering, science. Thorough analysis of theorems. Diagrams. Bibliography. Index. 818pp. 5⅜ x 8½.   0-486-64940-7

INTEGRAL EQUATIONS, F. G. Tricomi. Authoritative, well-written treatment of extremely useful mathematical tool with wide applications. Volterra Equations, Fredholm Equations, much more. Advanced undergraduate to graduate level. Exercises. Bibliography. 238pp. 5⅜ x 8½.   0-486-64828-1

FOURIER SERIES, Georgi P. Tolstov. Translated by Richard A. Silverman. A valuable addition to the literature on the subject, moving clearly from subject to subject and theorem to theorem. 107 problems, answers. 336pp. 5⅜ x 8½.   0-486-63317-9

INTRODUCTION TO MATHEMATICAL THINKING, Friedrich Waismann. Examinations of arithmetic, geometry, and theory of integers; rational and natural numbers; complete induction; limit and point of accumulation; remarkable curves; complex and hypercomplex numbers, more. 1959 ed. 27 figures. xii+260pp. 5⅜ x 8½.
0-486-63317-9

POPULAR LECTURES ON MATHEMATICAL LOGIC, Hao Wang. Noted logician's lucid treatment of historical developments, set theory, model theory, recursion theory and constructivism, proof theory, more. 3 appendixes. Bibliography. 1981 edition. ix + 283pp. 5⅜ x 8½.   0-486-67632-3

CALCULUS OF VARIATIONS, Robert Weinstock. Basic introduction covering isoperimetric problems, theory of elasticity, quantum mechanics, electrostatics, etc. Exercises throughout. 326pp. 5⅜ x 8½.   0-486-63069-2

THE CONTINUUM: A CRITICAL EXAMINATION OF THE FOUNDATION OF ANALYSIS, Hermann Weyl. Classic of 20th-century foundational research deals with the conceptual problem posed by the continuum. 156pp. 5⅜ x 8½.   0-486-67982-9

CHALLENGING MATHEMATICAL PROBLEMS WITH ELEMENTARY SOLUTIONS, A. M. Yaglom and I. M. Yaglom. Over 170 challenging problems on probability theory, combinatorial analysis, points and lines, topology, convex polygons, many other topics. Solutions. Total of 445pp. 5⅜ x 8½. Two-vol. set.
Vol. I: 0-486-65536-9 Vol. II: 0-486-65537-7

Paperbound unless otherwise indicated. Available at your book dealer, online at **www.doverpublications.com**, or by writing to Dept. GI, Dover Publications, Inc., 31 East 2nd Street, Mineola, NY 11501. For current price information or for free catalogues (please indicate field of interest), write to Dover Publications or log on to **www.doverpublications. com** and see every Dover book in print. Dover publishes more than 500 books each year on science, elementary and advanced mathematics, biology, music, art, literary history, social sciences, and other areas.

STATISTICAL PHYSICS, Gregory H. Wannier. Classic text combines thermodynamics, statistical mechanics and kinetic theory in one unified presentation of thermal physics. Problems with solutions. Bibliography. 532pp. 5⅜ x 8½. 0-486-65401-X

TENSOR CALCULUS, J.L. Synge and A. Schild. Widely used introductory text covers spaces and tensors, basic operations in Riemannian space, non-Riemannian spaces, etc. 324pp. 5⅜ x 8¼. 0-486-63612-7

ORDINARY DIFFERENTIAL EQUATIONS, Morris Tenenbaum and Harry Pollard. Exhaustive survey of ordinary differential equations for undergraduates in mathematics, engineering, science. Thorough analysis of theorems. Diagrams. Bibliography. Index. 818pp. 5⅜ x 8½. 0-486-64940-7

INTEGRAL EQUATIONS, F. G. Tricomi. Authoritative, well-written treatment of extremely useful mathematical tool with wide applications. Volterra Equations, Fredholm Equations, much more. Advanced undergraduate to graduate level. Exercises. Bibliography. 238pp. 5⅜ x 8½. 0-486-64828-1

FOURIER SERIES, Georgi P. Tolstov. Translated by Richard A. Silverman. A valuable addition to the literature on the subject, moving clearly from subject to subject and theorem to theorem. 107 problems, answers. 336pp. 5⅜ x 8½. 0-486-63317-9

INTRODUCTION TO MATHEMATICAL THINKING, Friedrich Waismann. Examinations of arithmetic, geometry, and theory of integers; rational and natural numbers; complete induction; limit and point of accumulation; remarkable curves; complex and hypercomplex numbers, more. 1959 ed. 27 figures. xii+260pp. 5⅜ x 8½. 0-486-63317-9

POPULAR LECTURES ON MATHEMATICAL LOGIC, Hao Wang. Noted logician's lucid treatment of historical developments, set theory, model theory, recursion theory and constructivism, proof theory, more. 3 appendixes. Bibliography. 1981 edition. ix + 283pp. 5⅜ x 8½. 0-486-67632-3

CALCULUS OF VARIATIONS, Robert Weinstock. Basic introduction covering isoperimetric problems, theory of elasticity, quantum mechanics, electrostatics, etc. Exercises throughout. 326pp. 5⅜ x 8½. 0-486-63069-2

THE CONTINUUM: A CRITICAL EXAMINATION OF THE FOUNDATION OF ANALYSIS, Hermann Weyl. Classic of 20th-century foundational research deals with the conceptual problem posed by the continuum. 156pp. 5⅜ x 8½. 0-486-67982-9

CHALLENGING MATHEMATICAL PROBLEMS WITH ELEMENTARY SOLUTIONS, A. M. Yaglom and I. M. Yaglom. Over 170 challenging problems on probability theory, combinatorial analysis, points and lines, topology, convex polygons, many other topics. Solutions. Total of 445pp. 5⅜ x 8½. Two-vol. set.
Vol. I: 0-486-65536-9 Vol. II: 0-486-65537-7

Paperbound unless otherwise indicated. Available at your book dealer, online at **www.doverpublications.com**, or by writing to Dept. GI, Dover Publications, Inc., 31 East 2nd Street, Mineola, NY 11501. For current price information or for free catalogues (please indicate field of interest), write to Dover Publications or log on to **www.doverpublications. com** and see every Dover book in print. Dover publishes more than 500 books each year on science, elementary and advanced mathematics, biology, music, art, literary history, social sciences, and other areas.